Applied Nanotechnology and Nanoscience International Conference (ANNIC 2016)

Journal of Physics: Conference Series Volume 829

Barcelona, Spain
9 - 11 November 2016

Editors:

Yian Tai
Sabherwal Priyanka

ISBN: 978-1-5108-3982-3
ISSN: 1742-6588

Printed from e-media with permission by:

Curran Associates, Inc.
57 Morehouse Lane
Red Hook, NY 12571

Some format issues inherent in the e-media version may also appear in this print version.

Printed by Curran Associates, Inc. (2017)

For permission requests, please contact the Institute of Physics
at the address below.

Institute of Physics
Dirac House, Temple Back
Bristol BS1 6BE UK

Phone: 44 1 17 929 7481
Fax: 44 1 17 920 0979

techtracking@iop.org

Additional copies of this publication are available from:

Curran Associates, Inc.
57 Morehouse Lane
Red Hook, NY 12571 USA
Phone: 845-758-0400
Fax: 845-758-2633
Email: curran@proceedings.com
Web: www.proceedings.com

Table of contents

Volume 829

Applied Nanotechnology and Nanoscience International Conference 2016 (ANNIC 2016)

9–11 November 2016, University Pompeu Fabra (Balmes building), Barcelona, Spain

Accepted papers received: 21 March 2017
Published online: 21 April 2017

Preface

Papers

IOP Conf. Series: Journal of Physics: Conf. Series **829** (2017) 011001 doi:10.1088/1742-6596/829/1/011001

Preface

Introduction:

The Applied Nanotechnology and Nanoscience International Conference is an annual event that hosts high-profile plenary speakers, world class researchers, oral and poster presentations, workshops, sponsor exhibits and extra activities to network.
This conference is held in a different European country each year and the number of participants is growing very quickly. ANNIC 2016 was held in Barcelona and hosted delegates from 47 different countries, it was a great opportunity for the attendees to share their research findings with wide audience, promote knowledge exchange and network.

Group Picture:

Main Sponsor: NFFA (nffa.eu)

Chairs Committee:

- Prof. Joan Ramon Morante (Catalonia Institute for Energy Research (IREC), Spain)
- Prof. Jordi Arbiol (ICREA & Catalan Institute of Nanoscience and Nanotechnology (ICN2), CSIC and The Barcelona Institute of Science and Technology (BIST), Catalonia, Spain)
- Prof. Arben Merkoçi (ICREA & Catalan Institute of Nanoscience and Nanotechnology (ICN2), Catalonia, Spain)
- Prof. Alberto Vomiero (Luleå University of Technology, Sweden)

Guest Editor Committee

- Dr. Yian Tai, National Taiwan University of Science and Technology, Taiwan
- Dr. Sabherwal Priyanka, Institute of Nano Science and Technology, India

Reviewers Committee:

- Laura M Lechuga, Spain
- Fernando Torres Andón, Italy
- Pablo Alonso-González, Spain
- Fabrice P. Laussy, Spain
- Toby Hallam, Ireland
- Robert S Maxwell, United States
- Olivier Soppera, France
- Bouraoui Ilahi, Saudi Arabia
- Thierry Baron, France
- Brent Wagner, United States
- Sergey A. Maksimenko, Belarus
- Luigi Sirleto, Italy
- Alexander Obraztsov, Russian Federation
- Maria Tchernycheva, France
- Daniel Granados, Spain
- Juan P. Martínez-Pastor, Spain
- Blas Garrido, Spain
- Felix Casanova, Spain
- Rongping Wang, Australia
- Daniele Sanvitto, Italy
- David Cox, United Kingdom
- Kadic Muamer, Germany
- Amitav Sanyal, Turkey
- Jan Honolka, Czech Republic
- Andrea Liscio, Italy
- Luigi Sangaletti, Italy
- Igor E Agranovski, Australia
- Valentine Novosad, United States
- Bart Partoens, Belgium
- Victor Ryzhii, Japan
- Maxim Ryzhii, Japan
- Bekir Aktas, Turkey
- Vladimir Baulin, Spain
- Mónica L. Fanarraga, Spain
- Luca Businaro, Italy
- Fernando Calle, Spain
- Thierry Conard, Belgium
- Antonio Tejeda Gala, France
- Fikret Yildiz, Turkey
- Devika Chithrani, Canada
- Hans Fecht, Germany
- Yoshinobu Baba, Japan

- Masami Furuuchi, Japan
- Nguyen T. K. Thanh, United Kingdom
- Muhammad Hassan Sayyad, Pakistan
- Maurizio Ferrari, Italy
- Elisabetta Comini, Italy
- Philippe Djemia, France
- Andrei Kanaev, France
- Jacques Jupille, France
- Mallikarjuna Nadagouda, United States
- Haibo Zhu, Saudi Arabia
- Rahul Mane, India
- Ildoo Chung, Korea, Republic Of
- Rogério Gaspar, Portugal
- Helena F Florindo, Portugal
- Karine Andrieux, France
- Nirali Gondaliya, India
- Kumar Magnash, France
- Nancy Ranxing Li, United States

IOP Conf. Series: Journal of Physics: Conf. Series **829** (2017) 011002 doi:10.1088/1742-6596/829/1/011002

Peer review statement

All papers published in this volume of *Journal of Physics: Conference Series* have been peer reviewed through processes administered by the proceedings Editors. Reviews were conducted by expert referees to the professional and scientific standards expected of a proceedings journal published by IOP Publishing.

IOP Conf. Series: Journal of Physics: Conf. Series **829** (2017) 012001 doi:10.1088/1742-6596/829/1/012001

Resistive switching in TiO_2 nanocolumn arrays electrochemically grown

M Marik, A Mozalev, J Hubalek and M Bendova

CEITEC – Central European Institute of Technology, Brno University of Technology, Purkynova 123, Brno 61200, Czech Republic

E-mail: marian.marik@ceitec.vutbr.cz

Abstract. Resistive switching in metal oxides, especially in TiO_2, has been intensively investigated for potential application in non-volatile memory microdevices. As one of the working mechanisms, a conducting filament consisting of a substoichiometric oxide phase is created within the oxide layer. With the aim of investigating the filament formation in spatially confined elements, we fabricate arrays of self-ordered TiO_2 nanocolumns by porous-anodic-alumina (PAA)-assisted anodizing, incorporate them into solid-state microdevices, study their electron transport properties, and reveal that this anodizing approach is suitable for growing TiO_2 nanostructures exhibiting resistive switching. The electrical properties and resistive switching behavior are both dependent on the electrolytic formation conditions, influencing the concentration and distribution of oxygen vacancies in the nanocolumn material during the film growth. Therefore, the PAA-assisted TiO_2 nanocolumn arrays can be considered as a platform for investigating various phenomena related to resistive switching in valve metal oxides at the nanoscale.

1. Introduction

Investigation of switching behavior of several metal oxides, which may have their resistance states modulated by voltage applied to or by the current flowing through them, is nowadays a growing research field motivating researchers and engineers to develop high-speed low-power nano-dimensional non-volatile Resistive Random Access Memory (ReRAM) devices. In such devices, a layer of metal oxide is usually sandwiched between two metallic electrodes for obtaining metal/semiconductor/metal structures. TiO_2, among other metal oxides like Ta_2O_5, HfO_2, ZrO_2, ZnO, WO_3, or NiO is one of the most intensively investigated resistive switching materials because of many advantages, such as the versatile fabrication techniques and suitable electrical properties resulting from a wide band gap and high dielectric constant [1,2,3].

A common mechanism of resistive switching in metal oxides is the formation of a thin conducting filament within the oxide layer, based on the field-induced movement of mobile oxygen vacancies [4,5]. A further filament rupture and re-creation at one of the interfaces between the metal oxide and a metal electrode, while applying voltage, is often the underlying phenomenon behind a bipolar resistive switching [5]. The filament is thus composed of a substoichiometric crystalline phase of the metal oxide, a Magnéli phase [6], formed also due to the high temperatures achieved during the local heating [7].

Nanostructuring of TiO_2 to the lateral size being comparable with a filament diameter (~15 nm [6]) may spatially confine the filament formation, leading to phenomena different from those occurring in continuous thin films. This motivated our efforts to incorporate arrays of TiO_2 nanocolumns prepared via the porous-anodic-alumina (PAA)-assisted anodizing [8,9] into a 3-D solid-state microdevice and investigate their electrical transport properties. The advantage is that the sizes of nanocolumns can be well controlled through the anodizing variables. In addition, the specific electrolytic conditions applied during the PAA-assisted film growth may enable tuning stoichiometry of the oxides and thus modification of their electrical behavior. In the present work, we fabricate PAA-assisted TiO_2 nanocolumn arrays, embedded in the anodic alumina matrix, by varying the anodizing conditions. We incorporate the arrays into metal/semiconductor/metal microdevices by electrochemically depositing top gold electrodes and measure their room-temperature current-voltage behavior, revealing the right combination of electrolytic conditions appropriate for the occurrence of bipolar resistive switching in the anodic films.

2. Experimental details

2.1. Fabrication of the TiO_2 arrays

A silicon wafer covered with a 500 nm thick thermal oxide layer was used as a starting substrate. A high quality Al/Ti bilayer (aluminum-on-titanium, 500/300 nm thick) was deposited via ion beam sputtering. The wafer with the deposited metal layers was cut into 10 mm ×15 mm pieces, which were individually processed in a closed two-electrode PTFE cell with a stainless steel counter electrode, exposing a surface area of 28 mm^2. Anodizing and re-anodizing experiments were performed in 0.3 M oxalic acid solution at 10 °C. Anodizing at a constant potential of 40 V led to formation of a PAA matrix having self-ordered nanopores, with nanoprotrusions of TiO_2 within the alumina barrier layer at the bottom of the PAA film having a mean diameter of ~35 nm. Re-anodizing to a higher anodic potential with a sweep rate of 2 V·s^{-1} resulted in relatively longer TiO_2 nanocolumns, while their length was tuned by setting the value of re-anodizing potential to 70, 100, and 130 V (arrays A) [10,9]. In addition, the nanocolumns grown to 100 V were prepared also by re-anodizing with a sweep rate of 0.2 V·s^{-1} (array B-100V). The typical current-time and potential-time transients obtained for the two sample types, re-anodized to 100 V, are shown in figure 1, supported by the schemes of nanoarray structure at the different stages of PAA-assisted film growth. The samples were further modified by annealing at 500 °C in vacuum (10^{-4} Pa).

To form the top contact to the TiO_2 nanocolumns and thus obtain a memory cell, gold was electrochemically deposited onto the column tops within and over the PAA nanopores [11,12]. The size of the top Au electrode was defined by an etch-back photolithography to ~100 μm × 100 μm.

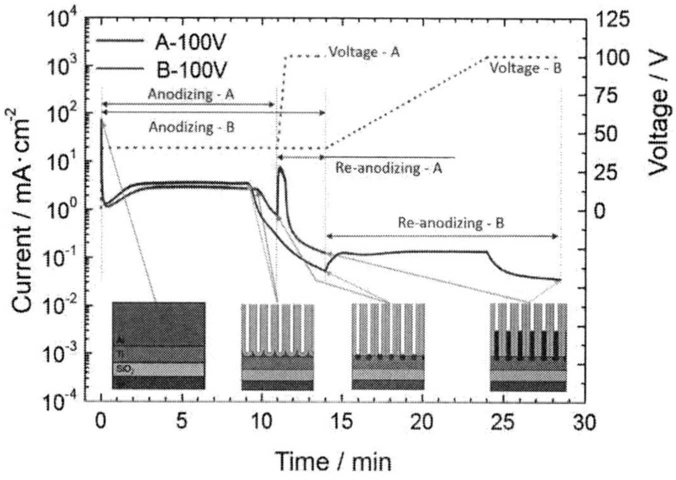

Figure 1. Current-time and potential-time transients recorded during the PAA-assisted anodization of arrays A-100V (red) and B-100V (black). The different regions (anodizing and re-anodizing) are marked and the images show schematically the arrays at the corresponding current or potential changes.

2.2. *Characterization*

Experimental samples (surfaces and cross-fractures) were observed by scanning electron microscopy (SEM) using a Tescan Mira II at different stages of sample fabrication. The electrical characterization was performed at room temperature using a Keithley 4200 SMU system connected to a CASCADE M150 probe station with magnetic micromanipulators, by recording $I(V)$ curves with a scan rate of 100 $mV \cdot s^{-1}$ from 0 to +5.0, then to −5.0 and back to 0 V in each cycle. To avoid damage of the microcells, the current compliance limit was set to 100 mA. The top Au pad was biased in all $I(V)$ measurements, while the bottom Ti layer, remaining after the PAA-assisted anodizing and being common for all devices prepared on a single chip, was grounded. The measured current and the calculated resistance were related to the projected area of the microcells.

3. Results and discussion

Arrays of TiO_2 nanocolumns are prepared via PAA-assisted anodizing to investigate their electrical properties with the focus on resistive switching application. The anodizing procedure is varied to obtain nanocolumns of variable length (increasing with the re-anodizing potential) and of different stoichiometry (possibly tuned by the potential sweep rate during the re-anodizing). To enable a good electrical contact to each nanocolumn in the array, gold is electrochemically deposited onto the TiO_2 columns, filling the alumina pores and growing over the alumina surface, which is constrained to square-shaped pads of a micrometer size. Typical examples of the devices and their cross fractures are displayed in the SEM images of figure 2, where the different array components can be seen and are schematically shown. A homogeneous filling of the alumina pores, with most of the TiO_2 nanocolumns contacted by gold, is desirable. The pore filling achieved in this study was not worse than 75%.

Typical initial $I(V)$ characteristics of the arrays A (i.e. having the *higher potential sweep rate* during re-anodizing) processed to different final potentials are shown in figure 3a in a logarithmic representation to emphasize the differences between the samples. Each of these arrays, comprising TiO_2 nanocolumns of various lengths, reveals non-linear shape of the corresponding $I(V)$ curve, also showing substantial current drops and jumps for the 70V and 130V arrays and an asymmetric diode-like behavior for the A-100V and A-130V samples. The initial resistance at 0 V ($R_{initial,0V}$) increases with increasing re-anodizing potential, from ~0.25 $\Omega \cdot cm^2$ for sample A-70V to ~7.5 $\Omega \cdot cm^2$ for both A-100V and A-130V arrays. Further $I(V)$ cycling (see figure 3b) shows again non-linear characteristics with an increased current density as compared with the initial state, being asymmetric only for the A-100V and A-130V arrays. The cycling resistance at 0 V ($R_{cycles,0V}$) is thus lower than that in the initial measurement by about ten and two times in case of sample A-70V and samples A-100V and A-130V respectively.

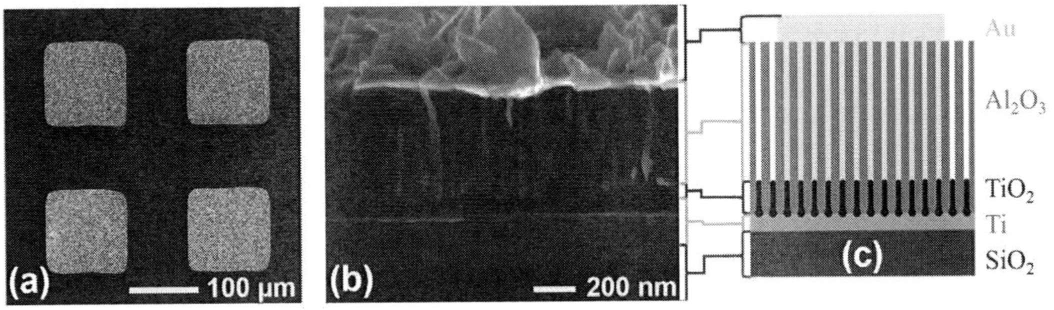

Figure 2. (a) Panoramatic and (b) cross-fractional SEM images of microcells consisting of A-100V nanocolumns. (c) Schematic drawing of the cross-section of a device, explaining the different layers visible in the SEM image of panel (b).

IOP Conf. Series: Journal of Physics: Conf. Series **829** (2017) 012001 doi:10.1088/1742-6596/829/1/012001

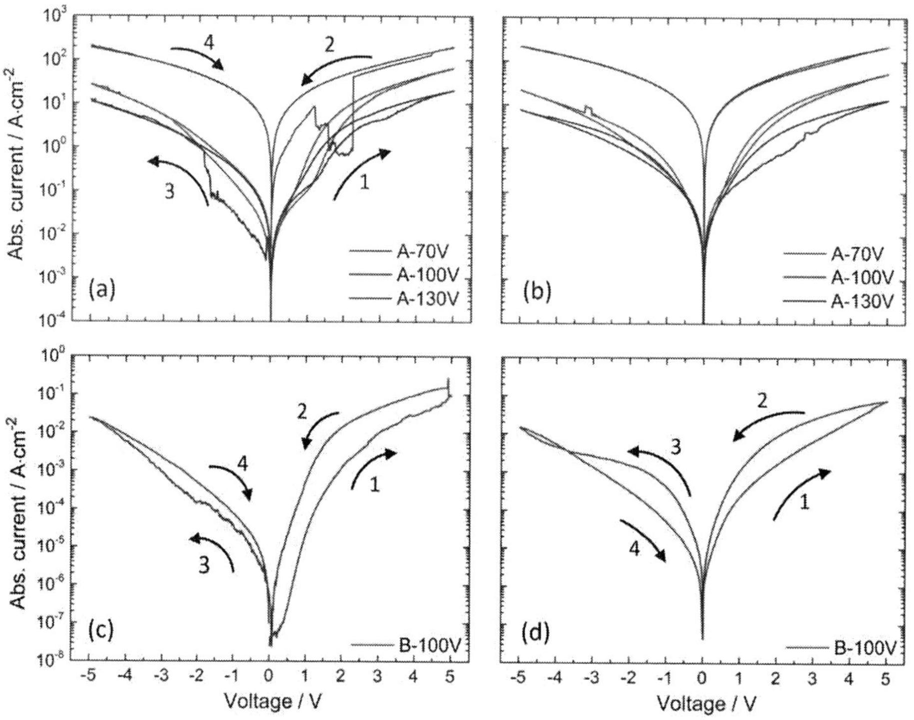

Figure 3. (a,c) The initial and (b,d) further $I(V)$ cycles obtained (a,b) for arrays A (i.e. formed at the higher potential sweep rate during re-anodizing) and (c,d) for arrays B (i.e. formed at the lower potential sweep rate during re-anodizing). The cycling was performed in the order 0 → +5.0 → −5.0 → 0 V without any break in applying voltage, at a scan rate of 100 $mV \cdot s^{-1}$ at room temperature. The top Au electrode was biased while the bottom Ti layer was grounded. The arrows indicate the scan direction.

The electrical characterization of the arrays A therefore does not reveal any resistive switching behavior, for which a substantial decrease of resistance after the initial $I(V)$ cycle accompanied by a presence of high- and low-resistance states (HRS and LRS, respectively) in the following cycles is expected [12]. The observed current jumps and drops in the initial $I(V)$ curves of arrays A (figure 3a) are rather caused by imperfect contacting, and there are no signs of distinguishable HRS and LRS in the following cycles. On the other hand, the resistance increases with the re-anodizing potential, which is in line with the increasing length of the nanocolumns and with their semiconductive nature. In addition, as samples A-100V and A-130V have their initial and consecutive $I(V)$'s similar to each other but different from those of sample A-70V (see the curve shape and symmetry), different conduction mechanisms are expected in these array groups. The oxygen-deficient root structure [13] of A-70V columns in the PAA barrier may prevail in the shorter nanocolumns, whereas a more stoichiometric, i.e. less conducting column body situated over the roots may lead to the diode-like behavior of A-100V and A-130V arrays, possibly with a Schottky barrier created at the Au/TiO_2 interface [12].

To alter the electrical behavior of the PAA-assisted TiO_2 nanocolumns, an array of type B (i.e. having the *lower potential sweep rate* during re-anodizing) was also prepared and electrically characterized. An initial $I(V)$ curve of sample B-100V (figure 3c) shows a non-linear asymmetric behavior with $R_{\text{initial,0V}}$ of ~$6 \cdot 10^5$ $\Omega \cdot cm^2$ and a current jump at 5.0 V, leading to a drop in the resistance of about one order of magnitude. The consecutive cycling (figure 3d) reveals also a non-linear asymmetric $I(V)$ characteristic and, in addition, the presence of HRS and LRS having $R_{cycles,0V}$ of ~$3 \cdot 10^4$ and ~$8 \cdot 10^3$ $\Omega \cdot cm^2$, respectively.

The array B-100V shows therefore signs of bipolar resistive switching. During the initial, forming $I(V)$ cycle, a conducting filament is probably formed within the columns or at one of the interfaces [12]. The switching direction, i.e. the transformation of HRS into LRS, happening at the positive polarization of the top Au electrode, followed by a transition from LRS into HRS at the negative polarization, indicates that the switching interface is the bottom one (TiO_2/Ti) [12]. The absence of abrupt current or resistance change during one of these transitions may indicate synaptic behavior of the array [14], executed in this case by inhomogeneous switching events in the multiple channels present in the array.

Comparing the anodizing conditions during the nanocolumn growth with the electrical behavior of arrays A and B, we see the following relations, summarized schematically in figure 4. The higher potential sweep rate during the re-anodizing (arrays A) leads to the formation of nanocolumns having a relatively low initial resistance (between 0.25 and 8 $\Omega \cdot cm^2$), whereas the lower potential sweep rate (array B) seems to increase the resistance ($6 \cdot 10^5$ $\Omega \cdot cm^2$). This may be explained by a formation of more oxygen-deficient and thus more electrically conducting nanocolumns in the case of the faster sweep rate during the column formation (arrays A) as compared to more stoichiometric oxide formed during the slower sweep rate (arrays B) [15]. Such higher-resistance, more-stoichiometric TiO_2 nanocolumns would be expected to lead to bipolar resistive switching behavior [5], as it is indeed observed in the present study. In addition, the length of the nanocolumns seems to have also an influence on their electrical resistance, possibly due to an inhomogeneous distribution of oxygen vacancies along the nanocolumn material, as manifested in arrays A by the increase in the resistance with increasing nanocolumn length, accompanied by the diode-like $I(V)$ characteristics of the longer A-100V and A-130V nanocolumns. To confirm this mechanism, as well as to elucidate the role of the annealing in vacuum on the concentration and distribution of oxygen vacancies, further analytical and electrical characterization of the arrays is being performed. Additionally, TiO_2 nanocolumn arrays grown at modified electrochemical conditions are currently under investigation; the results to be reported in due course.

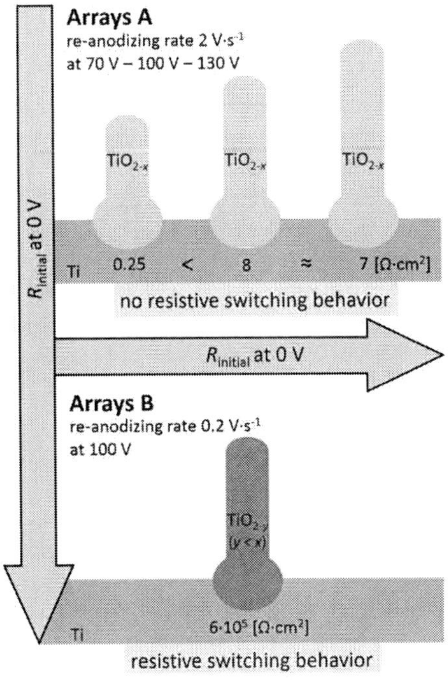

Figure 4. Schematic representations of PAA-assisted-anodized TiO_2 nanocolumns re-anodized to different potentials showing the relations between the anodizing conditions, oxide stoichiometry, and electrical properties. The faster re-anodizing rate (arrays A, upper part) results in a lower $R_{initial,0V}$ and reveals no sign of resistive switching, whereas the slower re-anodizing rate (arrays B, lower part) leads to the formation of more stoichiometric columns having a high $R_{initial,0V}$ and showing the resistive switching.

4. Conclusion

In this study, PAA-assisted well-ordered TiO_2 nanocolumn arrays were prepared and electrically characterized. Two different array types, A and B, grown at modified anodizing conditions, showed significant differences between their electrical and especially resistive switching behavior. Type A arrays, grown at a faster potential sweep rate, revealed lower resistances, increasing with the column length, and showed no sign of resistive switching. On the other hand, the nanocolumns fabricated at a slower potential sweep rate (type B) showed higher initial resistances and revealed a bipolar resistive switching behavior. The behaviors correlate with the oxygen-vacancy distribution established in the nanocolumns during anodic film growth. Thus, the PAA-assisted TiO_2 nanocolumn arrays form a basis for further development of resistive switching microdevices and for investigation of fundamental phenomena like the size-confinement effect in the metal-oxide nanostructures crucial for the filament formation.

Acknowledgements

Research leading to these results was supported in part by the Czech Science Foundation (GA ČR) under the grant no. 15-23005Y and in part by the Ministry of Education, Youth and Sports of the Czech Republic under the project CEITEC 2020 (LQ1601).

References

[1] Vazquez R M, Gispert-Guirado F, Llobet E and Mozalev A 2014 Proceedings of IEEE sensors 2014 (IEEE Service Center, USA) pp 807–810 DOI: 10.1109/ICSENS.2014.6985122
[2] Yang J J, Strukov D B and Stewart D R 2013 *Nat. Nanotechnol.* **8** 13–24
[3] Yang J J, Pickett M D, Li X, Ohlberg D A A, Stewart D R and Williams R S 2008 *Nat. Nanotechnol.* **3** 429–433
[4] Sawa A 2008 *Mater. Today* **11** 28–36
[5] Waser R, Dittmann R, Staikov G and Szot K 2009 *Adv. Mater.* **21** 2632–2663
[6] Kwon D-H, Kim K M, Jang J H, Jeon J M, Lee M H, Kim G H, Li X-S, Park G-S, Lee B, Han S, Kim M and Hwang C S 2010 *Nat. Nanotechnol.* **5** 148–153
[7] Marchewka A, Roesgen B, Skaja K, Du H, Jia C-L, Mayer J, Rana V, Waser R and Menzel S 2016 *Adv. Electron. Mater.* **2** 1500233
[8] Tatarenko N I and Mozalev A M 2001 *Solid-State Electron.* **45** 1009–1016
[9] Sjöström T, Fox N and Su B 2010 *Electrochim. Acta* **56** 203–210
[10] Vázquez R M, Calavia R, Llobet E, Guirado F, Hubálek J and Mozalev A 2012 *Procedia Eng.* **47** 833–836
[11] Mozalev A, Bendova M, Vazquez R M, Pytlicek Z, Llobet E and Hubalek J 2016 *Sens. Actuators, B* **229** 587–598
[12] Bendova M, Hubalek J and Mozalev A 2016 *Adv. Mater. Interfaces* **3** 1600512
[13] Chang Y-H, Lin H-W and Chen C 2012 *J. Electrochem. Soc.* **159** D512–D517
[14] Chang T, Jo S-H, Kim K-H, Sheridan P, Gaba S and Lu W 2011 *Appl. Phys. A* **102** 857–863
[15] Knörnschild G, Poznyak A A, Karoza A G and Mozalev A 2015 *Surf. Coat. Technol.* **275** 17–25

CVD-growth of MWCNT arrays on Me-Ct-N-(O) thin films

S Dubkov[1],A Trifonov[1,2], E Kitsyuk[3], A Pavlov[4], S Bulyarsky[4],S Skorik[3], T Maniecki[5], P Mierczynski[5], D Gromov[1], S Gavrilov[1]
[1] National research university of electronic technology, Zelenograd 124498, Russia
[2] Federal State Unitary Enterprise Lukin Research Institute of Physical Problems, Zelenograd 124460, Russia
[3] NPK Technological Center, Zelenograd 124498, Russia
[4]Institute of Nanotechnologies of Microelectronics, Russian Academy of Sciences, Moscow 115487, Russia
[5]Institute of General and Ecological Chemistry, Lodz University of Technology, 90-924 Lodz, Poland

sv.dubkov@gmail.com

Abstract. The formation of carbon nanotubes (CNTs) arrays by chemical vapor deposition on Me-Ct-N-(O) catalyst alloys, where Ct is a catalytic metal from the group Ni, Co, Fe, Pd; Me is transition metal of IV-VII of the Periodic Table of Elements, with low content of catalyst has been investigated. It is shown that CNTs effectively grow, if the alloy contains Ti, V, Cr, Zr, Hf, Nb and Ta. The structure of CNT were studied by transmission and scanning-electron microscopy, energy-dispersive X-ray and the Raman spectroscopy.

1. Introduction

Over the quarter century since the discovery of CNTs, their unique chemical, physical and mechanical properties have been well studied [1,2]. Due to these properties, CNTs are one of the most promising materials for use in various fields of science and technology [3-5].

Electric arc discharge, laser ablation of graphite and CVD are the most common methods of CNT synthesis[2]. From the viewpoint of reproducibility and growth process control, the chemical vapor deposition method is most suitable for formation of CNT array on substrate surface. This method implements the CNT growth process in the presence of catalysts such as Fe, Co, Ni, Pd and their alloys with each other, where the cluster size determines the parameters of the tube [6, 7]. It is known that the parameters of nanotube arrays may be controlled via the support of the catalytic metal by oxides of non-catalytic elements Al_2O_3, SiO_2, MgO [8, 9]. The authors [10,11] have reported about the use of the alloys of these catalysts with other metals, wherein the catalyst content is predominant. Yue Hu et al. in [12] have demonstrated an interesting method of catalyst preparation for the growth of single-walled CNTs. However, the method of preparation of such catalyst is rather complicated. In [13] it was shown that the array of multiwall carbon nanotubes can be formed on a thin film of catalyst-containing Nb-Ni-N-(O) amorphous alloy with low (~10 at.%) content of Ni. Distinctive feature of this process is that it can be performed on Nb-Ni-N-(O) alloy films of varying thickness, which makes it technically attractive.

In the present study we demonstrated that a similar process can be carried out on a number of alloys of catalyst with other transition metals V-VII of the Periodic Table of Elements. This process is dependent on a wide range of factors that allows you to control the parameters of CNT arrays, such as height, density and diameter.

2. Experimental

In this work were used (100) Si substrate, coated with SiO_2 layer. The substrates were washed instandard CARO solution (H_2SO_4:H_2O_2 = 1:1), and then rinsed in deionized water and dried in isopropanol vapor.

In alloy composition Me-Ct-N the element Ct was selected from the group Ni, Co, Fe, and Pd; element Me – from the group Ti, V, Cr, Zr, Nb, Mo, Hf, Ta, W and Re. Alloy layers with thickness from 10 to 500 nm with a catalytic metal content of 6–40 at.% and different nitrogen content were deposited on an unheated substrate by magnetron sputtering of a composite target Ct-Me in the atmosphere of argon or mixture of argon and nitrogen at a residual pressure in the chamber of $1 \cdot 10^{-5}$ Torr and a working gas pressure of $5 \cdot 10^{-3}$ Torr. Nitrogen content in the gas mixture Ar+N_2 varied in the range 0–10 vol.%.

CNTs were grown by CVD using Oxford PlasmaLab System 100 plasma etcher. CNT growth was carried out in CVD mode without plasma in the flow of acetylene with the addition of ammonia. Reactive gas flow for all the samples was constant. We varied the following parameters: synthesis temperature (in the range 400–750°C), synthesis time and the time of holding the sample at temperature of synthesis prior to supplying reactive gas.

CNTs samples were investigated using dual-beam scanning electron microscope FEI Helios NanoLab 650 i, transmission electron microscope FEI Tecnai G^2 20 S-Twin, equipped with EDAX energy dispersive X-ray spectrometer and HAADF detector, Horiba Scientific Lab RAM HR Evolution Raman spectrometer equipped with lasers with wavelength of 514.5 nm and 633 nm.

3. Results and discussion

Figure 1 shows the SEM image of CNTs grown on 20 nm thick film of the Zr-Co-N- (O) alloy. It should be noted that the array looks like rather dense forest of CNTs, about 9 μm in height.

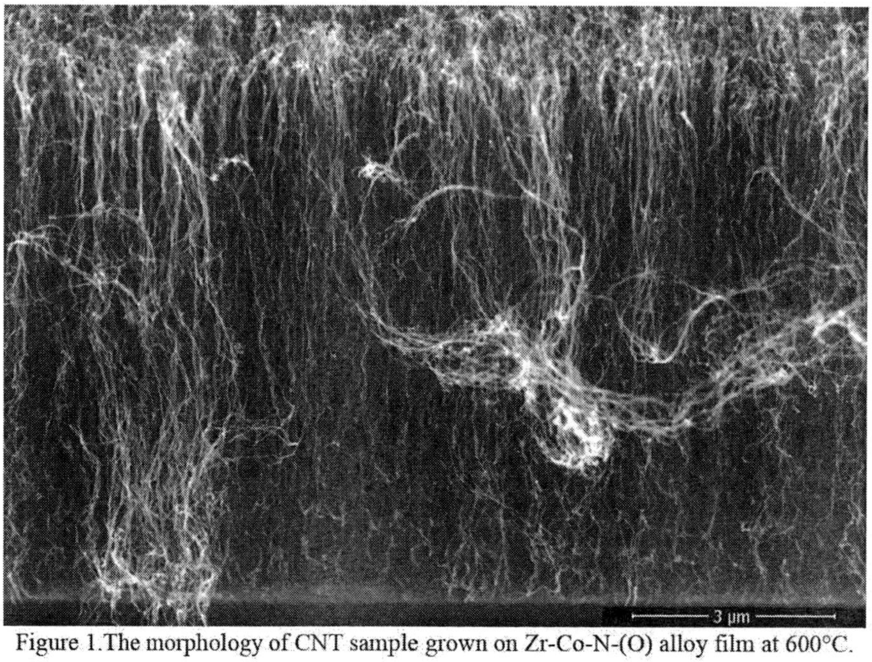

Figure 1.The morphology of CNT sample grown on Zr-Co-N-(O) alloy film at 600°C.

We found that CNT diameter depends on the metal Me, one of the components in the alloy. The thinnest CNTs were obtained on the alloys containing zirconium, while the thickest were synthesized on the vanadium alloy.It should be noted that CNT growth was not observed on W-Ni-N-(O), Re-Ni-N-(O), and Ta-Pd-N-(O) alloys. Instead free-shape carbon buildups were synthesized as shown in Figure 2. Size offree-shape buildupsranged from 50 to 200 nm.

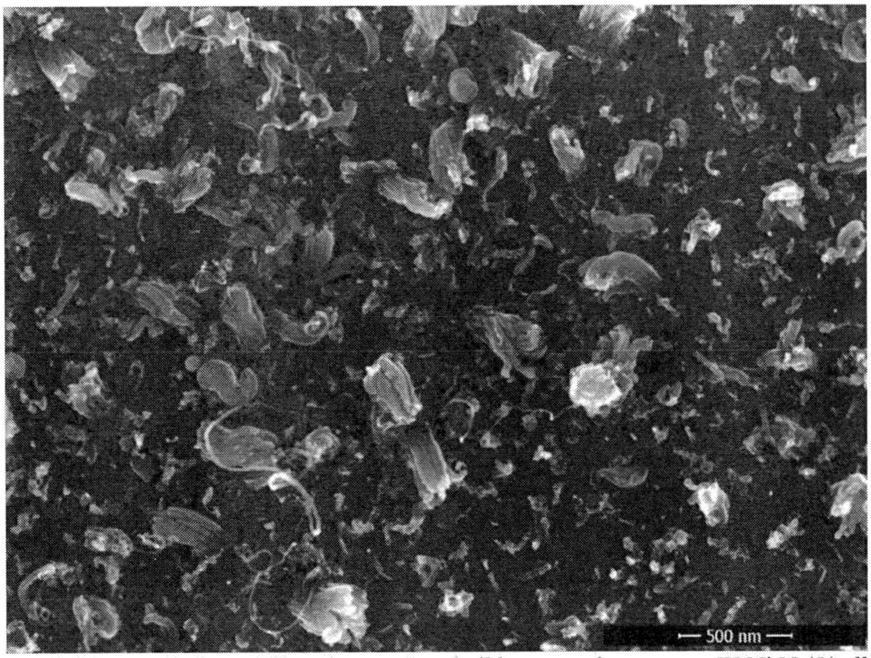

Figure 2. The morphology of free-shape carbon buildups sample grown on W-Ni-N-(O) alloy film at 650°C.

In the Raman spectrum of CNTs shown in Figure 2, D (1345 cm^{-1}), G (1580 cm^{-1}) and 2D (2700 cm^{-1}) peaks are present. The separation of D and G peaks and the ratio $I_D/I_G \sim 0.90$, indicate both good quality, and the presence of defects in the CNTs structure [14]. In turn, the presenceof intense 2D (or G') peak in spectrum indicates sufficient quality of carbon structure [15]. For other CNTs obtained on Me-Ct-N-(O) alloys, the positions and separation of the peaksin Raman spectra are substantially the same; minor changes were observed only in I_D/I_G ratio.

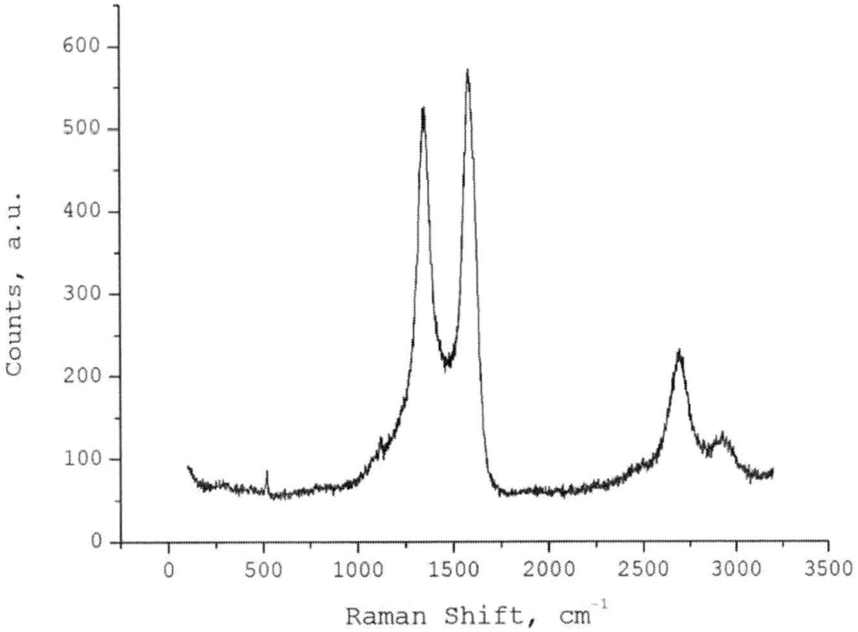

Figure 3. Raman spectra of CNT grown on a thin film of V-Ni-N-(O) alloy thickness of 20 nm at 600°C.

Figure 4 shows the results of investigation of nanotubes by transmission electron microscopy. On the high-resolution image we can see that the nanotubes are multiwall. Some of them have inclusions in the internal channels of CNTs. EDX spectrum of the inclusion is also shown in Fig. 4. From the spectrum we can conclude that this inclusions are the particles of the catalytic metal Ni.

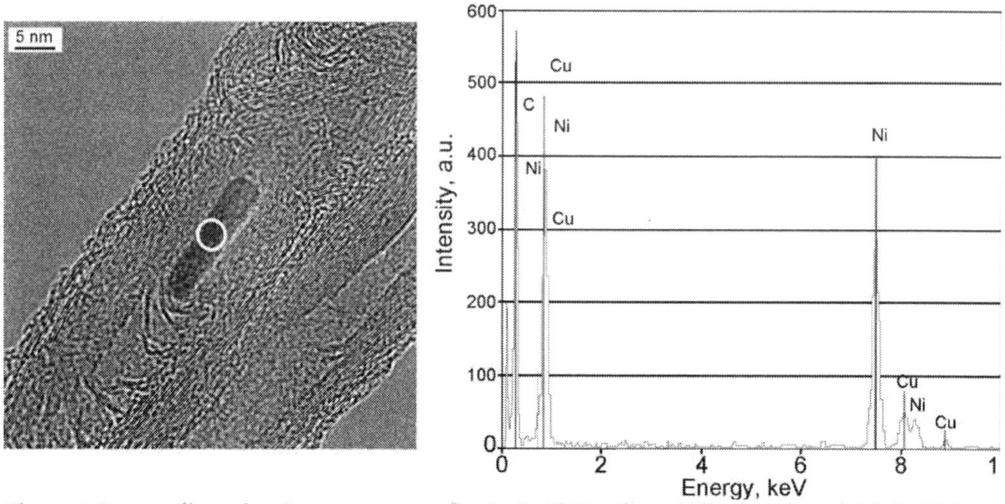

Figure 4. Energy dispersive X-ray spectrum of a single CNT with metallic inclusions (right). High-resolution TEM image of the object, where location of the analysis is marked by a circle. (left).

4. Conclusions

In commonly used methods of growth of carbon nanotubes, the catalyst particles are formed on the surface of a solid body through evaporation of extremely small amount of catalyst. This method is difficult to incorporate into integrated technology where the devices are formed on substrates of large diameters, since even slight variations of thickness cause substantial change of size and the number of clusters and therefore leads to the formation of CNTs differing in diameter and a height at different substrate areas.

The attractiveness of proposed process of CNT growth on the films of catalytic metal alloys with metals IV-VII of the Periodic Table of Elements is a broad set of factors for process control that allows you to manage the parameters of the nanotube arrays, such as height, density and diameter. Our experiments have shown that CNT growth is possible on thin alloy films of various thickness and conductivity. All this enables embedding of this process into integrated technology.

Acknowledgments

Russian Science Foundation financially supported this work (project number 16-19-10625).

References

[1] Iijima S. 1991 *Nature* **354**
[2] Dresselhaus M S, Dresselhaus G, Avouris P. 2001 *Springer-Verlag* **449**
[3] Kottler C *et al.* 2014 *Microelectronic Engineering* **122**
[4] Subash S, Kolar J & Chowdhury M H. 2013 *IEEE Transactions on Nanotechnology* **12**
[5] Xiao X *et al.* 2014 *Nano Energy* **6**
[6] Sohn J I, Ok Y-W, Seong T-Y & Lee S. 2007 *Journal of Applied Physics* **102**
[7] Liu H, Zhang Y, Li R, Sun X & Abou-Rachid H. 2011 *Particuology* **9**
[8] Bacsa R R *et al.* 2000 *Chemical Physics Letters* **323**
[9] Kitiyanan B, Alvarez W E, Harwell J H & Resasco D E. 2000 *Chemical Physics Letters* **317**
[10] Lobiak E V *et al.* 2015 *Journal of Alloys and Compounds* **621**
[11] Xu F, Zhao H & Tse S D. 2007 *Proceedings of the Combustion Institute* **31**

[12] Hu Y *et al* .2015 *Nature Communications* **6**
[13] Dubkov S *et al.* 2016 *Journal of Physics: Conference Series* **741**
[14] Vollebregt S, Derakhshandeh J, Ishihara R, Wu M Y & Beenakker C I. 2010 *Journal of Electronic Materials* **39**
[15] Antunes E F, Lobo A O, Corat E J & Trava-Airoldi V J. 2007 *Carbon* **45**

Diffusion of nanoparticles in solution through elastomeric membrane

Mohamed Zemzem*, **Ludwig Vinches**, **Stéphane Hallé**

Dept. of Mechanical Engineering, École de Technologie Supérieure
1100 Notre-Dame West, Montreal (QC) H3C 1K3 Canada

E-mail: mohamed.zemzem.1@ens.etsmtl.ca

Abstract. Diffusion phenomena encountered in mass transfer of liquids play an important role in many technological processes of polymer manufacturing and use. In addition and alongside the notable growth of nanoparticles use, particularly when in suspension in liquid solutions, it has become important to pay some attention to their interactions with polymeric structures. The aim of this work is to evaluate some diffusion parameters of gold nanoparticle solutions as well as of their liquid carrier (water) through elastomeric membranes. Gravimetric method was chosen as the main technique to quantify swelling phenomena and to assess kinetic properties. The dynamic liquid uptake measurements were conducted on gold nanoparticles (5 nm and 50 nm in diameter) in aqueous solutions when brought into contact with two types of nitrile material samples. Results showed that diffusion mechanism of the liquids lies between Fickian and sub-Fickian modes. Slight deviations were noticed with the gold nanoparticle solutions. A growth in liquid interaction with the rubbery structure in presence of the nanoparticles was also observed from comparison of K factor (characteristic of the elastomer-liquid interaction). Difference between the characteristics of the two membranes was also reported using this parameter. Besides, diffusion coefficients testified the impact of the membrane thickness on the penetration process, while no significant effect of the nature of the nanoparticle solution can be seen on this coefficient.

Keywords: Diffusion mechanism, gold nanoparticle solutions, nitrile rubber films, gravimetric method.

1. Introduction

The study of sorption and transport of liquids into polymeric systems is an important area of research that finds applications in many processes of industrial relevance [1]. Molecular transport of solutions through dense elastomeric films is found to obey the solution-diffusion model of transport in which, permeants dissolve in the membrane material and then diffuse through the membrane down a chemical potential gradient [2]. The diffusion of molecular species through a polymeric matrix of a membrane is affected by many factors such the liquid properties, the nature of the rubbery structure as well as the types of membrane-penetrant system interactions [3]. On the other hand, colloidal suspensions, particularly gold nanoparticle solutions are increasingly the subject of substantial research for their unique optical, electronic, and molecular-recognition properties [4]. However, there is a lack of information about the mechanisms of their transport through the polymeric structures. Therefore, the aim of this study is to develop a state of knowledge about some kinetic parameters of transport of gold nanoparticle solution through nitrile rubber membranes. The transport phenomenon is often accompanied by the swelling of the polymeric structure under the effect of the chain's relaxation [5].

Hence, the liquid diffusion phenomenon and the processes of material inflation are closely interrelated. Quantitative data of swelling yield valuable information about transport mechanisms [6, 7]. Among several experimental techniques, gravimetric method was adopted in this study to assess swelling of nitrile films brought into contact with two gold nanoparticle solutions of different diameters.

2. Experimental

2.1. Materials

As elastomeric membrane, two common disposable nitrile rubber gloves were used for this study. Both had similar chemical composition but different thicknesses (73.2 ± 3.0 μm and 117.0 ± 6.5 μm). Until the end of this manuscript, both models are labeled as NBR1 and NBR2 respectively.

Regarding nanoparticle suspensions, two commercial gold nanoparticle solutions (5 and 50 nm) in MilliQ water were selected. Both suspensions had the same concentration of 0.05 mg/mL. To ensure their stability, the gold nanoparticles were coated with polyvinylpyrrolidone. Nanoparticle solutions were purchased from NanoComposix (San Diego, CA, USA), and they were labeled as AuNP-5 and AuNP-50 respectively.

2.2. Methods

Rectangular samples (60 × 10 mm) were taken from palm section of the protective gloves. Each sample was weighted using digital balance (±0.1 mg). At ambient temperature, samples were placed in glass bottles containing about 10 mL of gold nanoparticle solution. At regular periods of time, specimens were removed and were gently dried using a filter paper from any adhering liquid and weighted. The tests ended when the weight of the sample reached a plateau (equilibrium state).

Mass uptake data (M_t/M_∞) were represented as a function of the root square of time normalized by the thickness of the sample using the following equations [8]:

$$M_t/M_\infty = (m_t - m_0)/(m_\infty - m_0) \qquad (1)$$

Where, m_0 is the initial weight of sample, m_t is the weight at time t and m_∞ is the weight at the equilibrium.

2.3. Theories

Diffusion mechanism:

In order to investigate the type of transport phenomenon and the diffusion mechanism, the dynamic sorption data of all the penetrant-membrane systems have been fitted to the following empirical relation [9]:

$$\log(M_t/M_\infty) = \log K + n \log t \qquad (2)$$

Where, n is an index related to the type of the transport mechanism of the penetrant through the membrane structure. Regarding rubbery materials and liquid systems, the values of n are usually located between 0.5 and1[10]. For n equal to 0.5, the diffusion follows the Fickian trend in which the rate of the elastomer chains relaxation exceeds the rate of the liquid diffusion. On the other hand, for n equal to unity, the diffusion mode is non-Fickian and the diffusion rate of the permeant molecules is higher than the elastomer segment mobility. K factor depends upon the elastomer-liquid interaction as well as the structural features of the rubbery material [11].

Diffusion coefficients:

With Fickian transport mode, diffusion coefficients of an unidirectional flow of solution through the nitrile membranes can be calculated using the thin-film solution for the second Fick's law of mass transfer given by [12]:

$$\partial C/\partial t = D(\partial^2 C/\partial x^2) \tag{3}$$

Where, C refers to the liquid concentration within the nitrile membrane at a given time t and for a depth x inside the sample, D is the diffusion coefficient of the penetrant through the sample.

An analytical solution of this differential equation defines the liquid concentration during sorption through a given rubber sheet of thickness h, at time t, and a distance x. Equation (3) was solved under suitable initial and boundary conditions using some assumptions; the concentration of the colloidal solution on the membrane surface is supposed to reach equilibrium immediately upon exposure during the immersion test [13]:

$$\begin{array}{lll} for\ t = 0 & 0 < x < h & C = 0 \\ for\ t \geq 0 & x = 0, x = h & C = C_\infty \end{array} \tag{4}$$

$$C_{(t,x)}/C_\infty = 1 - \frac{4}{\pi}\sum_{n=0}^{\infty}\frac{1}{(2n+1)}\exp\left(\frac{-(2n+1)^2\pi^2 D}{h^2}t\right) \times \sin\left(\frac{(2n+1)\pi}{h}x\right) \tag{5}$$

Where $C_{(t,x)}$ is the concentration of the solution at time t and distance x from the membrane surface. C_∞ is the solution concentration at equilibrium and h refers to the thickness of the membrane. Integrating Equation (5) gives the mass of the absorbed penetrant by the film as a function of time. Under these conditions, mass uptake of the penetrant solution can be described by either of the following equations [14]:

$$M_t/M_\infty = 4\left(\frac{D}{\pi}\right)^{0.5}\cdot\left(\frac{\sqrt{t}}{h}\right) \tag{6}$$

$$\ln(1 - M_t/M_\infty) = \ln\left(\frac{8}{\pi^2}\right) - \pi^2 D\cdot\left(\frac{t}{h^2}\right) \tag{7}$$

Equation (6) converges rapidly at short times (where $M_t/M_\infty \leq 0.6$), whereas equation (7) converges at long times of immersion (where $M_t/M_\infty \geq 0.6$). Both expressions were limited to the first order of iteration (n=0), the error caused by the simplification is less than 0.1% [15].

Concentration profiles:
The determined values of D were incorporated into Fick's law solution equation (5) to generate theoretical sorption curves. Concentration profiles of the solutions at different depths of the nitrile films have been calculated.

3. Results and discussion

3.1. Mass uptake and elongation phenomena

Swelling takes place as liquid is adsorbed. This phenomenon depends upon the solubility characteristics of the elastomer-solution system. At a prolonged period of exposure, the penetrant causes a relaxation of the polymeric chains giving rise to more space to occupy by the liquid. The main change noticed after immersion is the mass uptake and the volume change of the rubbery sample [16]. Figure 1 represents mass uptake over time of NBR2 in contact with MilliQ water, AuNP-5 and AuNP-50. Results showed typical curves of liquid sorption by an amorphous polymeric structure. A gradual increase was noticed until attaining a state of saturation. The saturation was achieved after about 200 hours of immersion. At equilibrium, the nitrile rubber sample gained around 40% of its initial weight with MilliQ water. However, this gain was around 50% with each one of the nanoparticle solutions.

In order to be able to compare swelling behavior of the two materials, reduced sorption data plots were presented in Figure 2 as function of normalized time according to equation (1). Both curves of NBR1 and NBR2 in AuNP-50 were similar in form and exhibited the Fickian feature of linearity up to

IOP Conf. Series: Journal of Physics: Conf. Series **829** (2017) 012003 doi:10.1088/1742-6596/829/1/012003

about 60% of fractional weight uptake. When comparing the linear parts of the two nitrile films, it is worth noting that the slope the NBR2 curve seems to be slightly higher. It was also remarkable that NBR2 achieved saturation much earlier than NBR1.

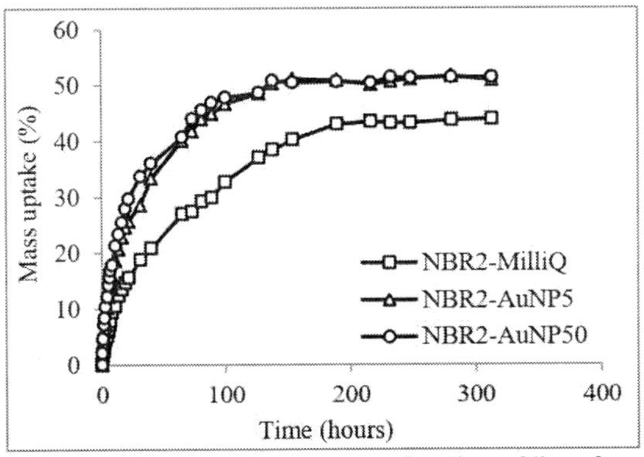

Figure 1 Mass uptake curves as function of time for NBR2 in MilliQ water, AuNP-5 and AuNP-50

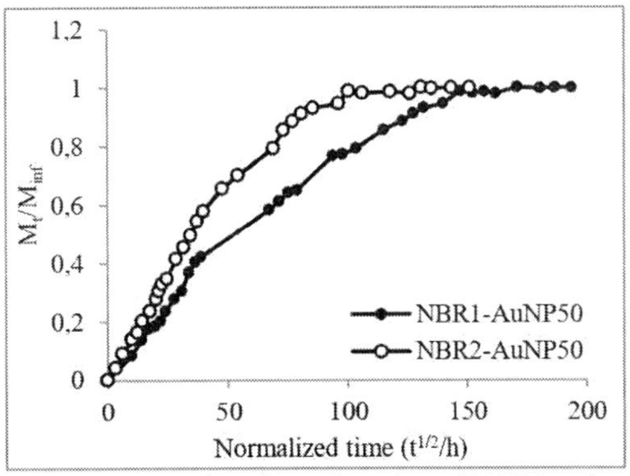

Figure 2. Reduced sorption plots of NBR1 and NBR2 in contact with AuNP-50 (M_{inf} refers to M_∞)

3.2. Kinetic parameters of diffusion

Power law model expressed under logarithmic form by equation (2) is reported to be the easiest and most common mean to describe sorption results [17]. Using fitting module of MATLAB software, experimental data were fitted to power law curve. Information about the diffusion mechanism of NBR1 and NBR2 in MilliQ water as well as in both nanoparticle solutions are represented in Table 1. The values of n vary from 0.42 to 0.55 for NBR1 and from 0.5 to 0.52 for NBR2 suggesting a slight deviation from Fickian diffusion (accurate to ±0.01 unit). As a side note, the size of the nanoparticles seems not having a significant effect on the diffusion mechanism for both nitrile rubber films. However, a related point to consider is that n index for NBR1 is 0.05 unit lower than the n index of NBR2, and that for both nanoparticle solutions. This fact is related to the difference in thickness of the two films.

Regarding K factor, there are two noteworthy aspects to consider: Firstly, and for all the liquids, a difference of about five orders of magnitude was noticed between the two membranes. A higher interaction of NBR1 with the solutions can be deduced. This fact may be due to a difference of the chemical composition of the two glove materials. Secondly, and for both films, K factors for nanoparticle solutions are from 2.5 to 3 times higher than the MilliQ water factors. This result highlights a possible effect of the nanoparticles on the membrane-penetrant interaction.

Table 1 Diffusion mechanism index, K factor and first order factor k_1 of the penetration of MilliQ water, AuNP-5 and AuNP-50 through NBR1 and NBR2

	NBR1		NBR2	
	n	K ($g \cdot g^{-1} \cdot min^{-n}$)	n	K ($g \cdot g^{-1} \cdot min^{-n}$)
MilliQ water	0.5572	9.736	0.5236	2.962
AuNP-5	0.4452	25.680	0.5056	5.369
AuNP-50	0.4276	31.01	0.5156	6.115

Sorption data (M_t/M_∞) curves were used to determine diffusion coefficients through the relations given by equation (6) and equation (7). Taking into account the change in the mass uptake profile over time, diffusion coefficients should be separately assessed for short and long times. Table 2 displays diffusion coefficients data, at short and long times, of MilliQ water, AuNP-5 and AuNP-50 through NBR1 and NBR2. First of all, it has not escaped our notice that for gold solutions, and less for MilliQ water, the thickness of the membrane is a very important parameter affecting the diffusion coefficients

[10], which also seen in equations (6) and (7). For NBR2 and the both gold suspensions and at short times, D was at least three times higher than the coefficients of NBR1. Nonetheless, no significant effect can be reported regarding the nature of the colloidal solution. In addition, and for NBR2, short diffusion coefficients of the nanoparticle solutions were notably higher than the coefficient of MilliQ water with the same material. This behavior was not seen with NBR1. This effect is consistent with the diffusion mechanism results and may be caused again by a different chemical composition between the two nitrile rubber materials. Moreover, when comparing diffusion coefficients using previously described equations at short and long times of immersion, insignificant differences were mentioned. Therefore, it can be deduced the continuity of the short and long models along the total immersion time.

Table 2 Diffusion coefficients ($\times 10^{-11}$ $cm^2 \cdot s^{-1}$) of MilliQ water, AuNP-5 and AuNP-50 through NBR1 and NBR2 for short and long times at first order of iteration (simplified expression).

	NBR1		NBR2	
	Short times	Long times	Short times	Long times
MilliQ water	4.13	3.96	4.87	5.23
AuNP-5	3.39	3.98	8.27	8.17
AuNP-50	3.33	4.11	10.29	8.95

A graphical representation of the liquid presence inside the polymeric structure was made possible due to equation (5). Fractional concentration of MilliQ water, AuNP-5 and AuNP-50 were plotted all long NBR2 membrane thickness at a fixed immersion time lapse of 50 hours (Figure 3). When the nitrile rubber sample is immersed, the liquid penetrates the structure until reaching the center of the film at a given time t. The concentration (C_t) increases in the core up to the initial concentration (C_t/C_∞=1). It is worth mentioning that the sample absorbed about 15% of MilliQ water in the center. However, and at the same immersion time (50 hours), the concentration ratio of AuNP-5 and AuNP-50 were between 60 and 70% respectively. Membrane structure seems to absorb more nanoparticle solutions than MilliQ water. An important related point to consider was that the colloidal solutions contain additives used as stabilizers for the dispersion of the nanoparticles. These substances may affect considerably the penetration process and be a part of the differences seen in the concentration profiles.

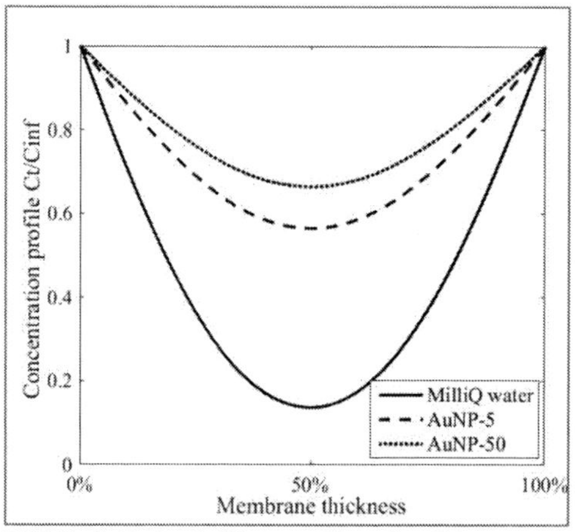

Figure 3 Concentration profiles of MilliQ water, AuNP-5 and AuNP-50 into the structure of NBR2 at 50 hours of immersion (C_{inf} refers to C_∞)

4. Conclusions

Research on sorption and diffusion properties of liquids through polymeric structures was extensively covered, along decades, by numerous studies. In spite of all that, the penetration of nanoparticle solutions through elastomeric membranes has not been well studied yet. Gravimetric method was retained as an effective technique to study penetration kinetics of the nanoparticle solutions through the nitrile rubber membrane structures. This paper presents an analysis of some penetration parameters, through nitrile membranes, of two gold nanoparticle solutions (5 and 50 nm of diameters) as well as of their liquid carrier (MilliQ water). Findings demonstrate that mechanism diffusion of the liquids ranges from sub-Fickian to Fickian diffusion mode. The noticed mode deviation can be explained by the presence of the gold nanoparticles in the solutions. An increase in liquid interaction with the rubbery structure in presence of the nanoparticles was also observed from comparison of K factor. Difference between the characteristics of the two membranes was also reported using this parameter. In addition, diffusion coefficients showed the effect of the membrane thickness on the penetration process, while no significant difference can be seen on this coefficient when adopting a short or long time model. As summary, presence of nanoparticles on the solutions can affect penetration phenomena. Further investigation is required to evaluate the kinetic parameters of these nanoparticles on the process.

References

[1] Chen, M.-J., et al., European Polymer Journal, 2015. **66**: p. 407-418.
[2] Wijmans, J.G. and R.W. Baker, Materials science of membranes for gas and vapor separation, 2006: p. 159-190.
[3] George, S.C. and S. Thomas, Progress in Polymer Science, 2001. **26**(6): p. 985-1017.
[4] Kumar, A., X. Zhang, and X.-J. Liang, Biotechnology Advances, 2013. **31**(5): p. 593-606.
[5] Lv, X.R., H.M. Wang, and S.J. Wang, Wear, 2015. **328–329**: p. 414-421.
[6] Vinches, L., et al., Journal ISSN, 2012. **1929**: p. 1248.
[7] Perron, G., J.E. Desnoyers, and J. Lara, Journal of applied polymer science, 2002. **86**(1): p. 195-215.
[8] *The sorption and diffusion of small molecules in amorphous and crystalline polybutadienes*. in *Journal of Polymer Science: Polymer Symposia*. 1973. Wiley Online Library.
[9] Aminabhavi, T.M. and R.S. Khinnavar, Polymer, 1993. **34**(5): p. 1006-1018.
[10] *Diffusion in polymer solids and solutions*. 2011: INTECH Open Access Publisher.
[11] Unnikrishnan, G. and S. Thomas, Journal of Polymer Science Part B: Polymer Physics, 1997. **35**(5): p. 725-734.
[12] Kwan, K.S., C.N. Subramaniam, and T.C. Ward, Polymer, 2003. **44**(10): p. 3061-3069.
[13] Aminabhavi, T.M., et al., Polymer, 1996. **37**(9): p. 1677-1684.
[14] Balik, C., Macromolecules, 1996. **29**(8): p. 3025-3029.
[15] *Liquid transport processes in polymeric materials: modeling and industrial applications*. 1991: Prentice Hall.
[16] George, S.C., M. knörgen, and S. Thomas, Journal of Membrane Science, 1999. **163**(1): p. 1-17.
[17] Mathai, A.E., R. Singh, and S. Thomas, Journal of membrane science, 2002. **202**(1): p. 35-54.

IOP Conf. Series: Journal of Physics: Conf. Series **829** (2017) 012004 doi:10.1088/1742-6596/829/1/012004

Pseudo-Hall Effect in Graphite on Paper Based Four Terminal Devices for Stress Sensing Applications

Afzaal Qamar[1], Tuba Sarwar[2], Toan Dinh[1], A.R.M. Foisal[1], Hoang-Phuong Phan[1] and Dzung Viet Dao[1,3]

[1] Queensland Micro-Nanotechnology Centre, Griffith University, Queensland, Australia
[2] DPAM, PIEAS, P. O. Nilore, Islamabad, Pakistan
[3] School of Engineering, Griffith University, Queensland, Australia

Abstract. A cost effective and easy to fabricate stress sensor based on pseudo-Hall effect in Graphite on Paper (GOP) has been presented in this article. The four terminal devices were developed by pencil drawing with hand on to the paper substrate. The stress was applied to the paper containing four terminal devices with the input current applied at two terminals and the offset voltage observed at other two terminals called pseudo-Hall effect. The GOP stress sensor showed significant response to the applied stress which was smooth and linear. These results showed that the pseudo-Hall effect in GOP based four terminal devices can be used for cost effective, flexible and easy to make stress, strain or force sensors.

1. Introduction

In recent years, researchers have paid much attention to paper based devices for diagnostics, electronics and microfluidics applications [1-3]. Flexible and wearable electronics is the fundamental part of flexible appliances which require ultra-thin and flexible navigation modules, force sensing, body tracking, and relative position monitoring systems. Soft robotics and transient electronics are the ultimate beneficiaries of these flexible and wearable electronics. The main advantages of this technology are the low cost, diversity in material choice, disposability and ease of fabrication [4]. Graphite on paper (GOP), in which graphite layers are deposited on paper by either graphite-ink printing techniques [5-6], or manual pencil drawing techniques, has been utilized as the sensing element in flexible sensors for strain sensing and chemical analysis [7-12].

Most of the previous work on GOP based stress/strain sensors is based on the conventional two terminal resistor sensors which suffer several drawbacks. They show high temperature sensitivity relative to stress response. To increase the total resistance, resistors are often designed with large meandering patterns which suffer from transverse sensitivity [13]. On the other hand pseudo-Hall effect device inherently utilizing the high accuracy of four-wire resistance measurement method can overcome the limitations of the conventional resistor sensors which show less temperature sensitivity and can be made small enough to capture the localized stress [14]. It can also eliminate the use of Wheatstone bridge for stress measurement which is used in simple resistor based strain sensors. A large number of studies can be found on Si based pseudo-Hall devices and SiC based pseudo-Hall device [15-21] but according to our knowledge there has been no report on the pseudo Hall effect in Graphite on paper based devices. Therefore, in this study we report a GOP based cost effective and flexible stress/strain sensor based on pseudo-Hall effect.

IOP Conf. Series: Journal of Physics: Conf. Series **829** (2017) 012004 doi:10.1088/1742-6596/829/1/012004

2. Characterization of the Graphite on Paper

In order to investigate the properties of the Graphite drawn on paper, scanning electron microscopy and Raman spectroscopy were used. Figure 1(a) shows the SEM image of the paper substrate. It can be observed that the paper substrate consists of cellulose fibre and is porous which helps the graphite particles to imbed in it. Figure 1(b) shows the SEM image of the Graphite trace on paper. It can be observed from Fig. 1(b) that a continuous and uniform trace of Graphite was obtained. In order to analyse the carbon networks and defect density inside the pencil trace Raman spectroscopy was utilized. Figure 1(c) shows the Raman spectra of the GOP trace. There are three main peaks observed in the Raman spectra corresponding to three different bands D, G and 2D at waver numbers of 1350 cm-1, 1580 cm-1 and 2725 cm-1, respectively. The D band corresponds to the number of defects and boundaries, while the G band indicates the number of sp2 bonded carbon networks in the Graphite trace. As the Graphite trace is made by pencil drawing the presence of the defects and boundaries is obvious by this technique but the enough carbon networks are present inside the trace for the establishment of current flow. Additionally, the 2D band corresponds to the type of stacking sequence in a direction perpendicular to the graphite plane. Figure 1(d) shows the photograph of a four terminal device (500 μm × 500 μm) made from pencil trace and after making the contacts by silver epoxy.

3. Fabrication of Devices

Figure 2 shows the fabrication process of the device. A simple pencil with 6B grade was used to draw Graphite trace on to a paper substrate (Fig. 2(a)). After the pencil trace low resistive aluminium tape and silver epoxy (186–3616, RSC components) were used to make the electrical contacts to the devices as shown in Fig. 2(b). The fabricated devices with silver epoxy contacts were then cured on a hotplate (Fig. 2(c)). The paper substrate was then cut into stripes to apply stress for pseudo Hall effect measurements (Fig. 2(d). The Ohmic behaviour of the fabricated devices was analysed using HP 4145B analyser, indicating a good Ohmic contact between the electrodes and the graphite trace. The input current to the device was applied at terminals 1 and 2 while the output voltage was observed at terminals 3 and 4 as indicated in Fig. 1(d). The contacts for input current was made strip like in order to have the uniform effect of stress in the device while the output contacts were made point like to have the minimum deviation in the output signal. Four terminal devices with different geometries were made and analysed for pseudo-hall effect including, square shape, rectangular shape and Hall cross shape but the square shape is presented here as it produced the maximum output response to stress.

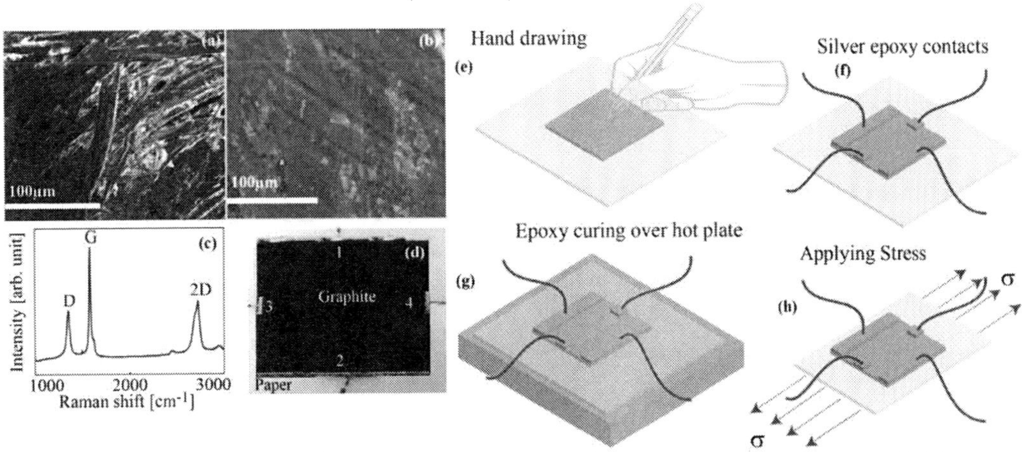

Figure 1: (a) SEM image of paper substrate, (b) SEM image of GOP drawn by pencil, (c) Raman spectra of GOP, (d) GOP four terminal device made from pencil drawing and contacts from silver epoxy, (e) schematic of the pencil drawing on paper, (f) fabrication of electrical contacts by silver epoxy, (g) curing of the epoxy to get stable Ohmic contacts, (h) cutting of the paper strip containing devices into strips to apply stress.

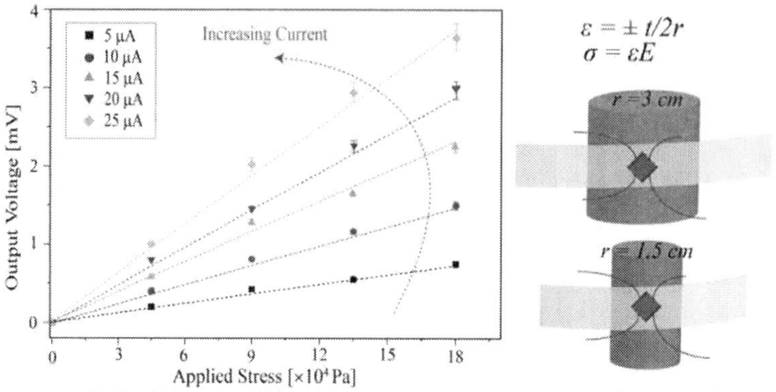

Figure 2: Pseudo-Hall effect in GOP square device. The output voltage of the device varies significantly as the stress is varied. The output voltage also increases linearly with increasing the value of input current. The stress was applied to the device by wooden cylinders having different diameters.

4. Results and discussion

The stress to the devices was applied by bending the paper substrate, containing graphite device, around wooden cylinders of different diameters. Different diameters of the cylinder produced different strain to the graphite devices according the relation; $\varepsilon = t/2r$. Here, t is thickness of the paper substrate, which is 90 μm and r is the radius of the wooden cylinder. The correspond stress induced to the devices was calculated by Hooke's law; $\sigma = \varepsilon E$, where E is the Young's modulus of the paper having a value of 0.03 GPa [22]. When stress is applied to the device at a constant input current at terminals 1 and 2, a variation in output voltage at terminals 3 and 4 can be observed, which is called pseudo-Hall effect. Figure 2(a) shows the variation of output voltage of the device against various levels of applied stress to the device. It can be observed that the output voltage of the device varies significantly when the applied stress to the device is varied and this variation is linear against the applied stress. This shows that the pseudo-Hall effect in GOP based devices can be used as a measure of stress/strain, force and pressure.

The output response of the device was observed at different angles within the plane of the paper. Figure 3 shows the variation of output voltage of the device against applied stress when the device is rotated within the plane of the paper. The square device was fabricated at different angles with respect to the plane of the paper. The output response of the device was different for different angles within the plane of the paper. The maximum output response was only observed when the direction of input current was at an angle of 45° with the direction of applied stress. The minimum out response was observed when the angle between direction of current and applied stress was 0° or 90°. Therefore, the output voltage against applied stress varies between maximum and minimum values as the device is rotated in the plane of paper.

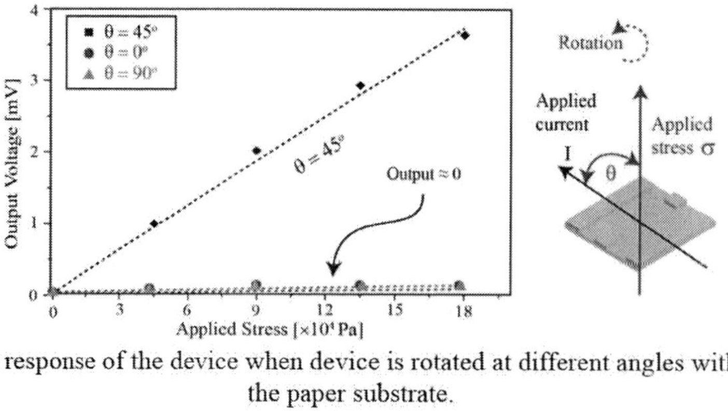

Figure 3: Output response of the device when device is rotated at different angles within the plane of the paper substrate.

In order to analyse the repeatability of the output response of the device against applied stress, the stress was applied to the device in steps. Figure 4 (a) shows the output response of the device when stress was applied and removed in steps of 10 sec. It can be observed that the response of the device to applied stress is repeatable and comes to initial value after each step when stress is removed. Figure 4(b) shows the variation of output voltage of the device against input current at a fixed value of the external applied stress. It can be observed from Fig. 4(b) that the output response of the device also increases linearly when the input current is increased. This feature of the device is very important as the output signal of the device can be raised to any measurable level by just increasing the input current to the device. This is not true in case of a simple piezoresistor in which a resistor produces change in resistance which is same at all input current values. The compressive stress/strain behaviour of these devices could not be observed due to the damage of the device when it is compressed, facing the wooden roller. This problem can be eliminated by laminating the paper after making the device and ohmic contacts.

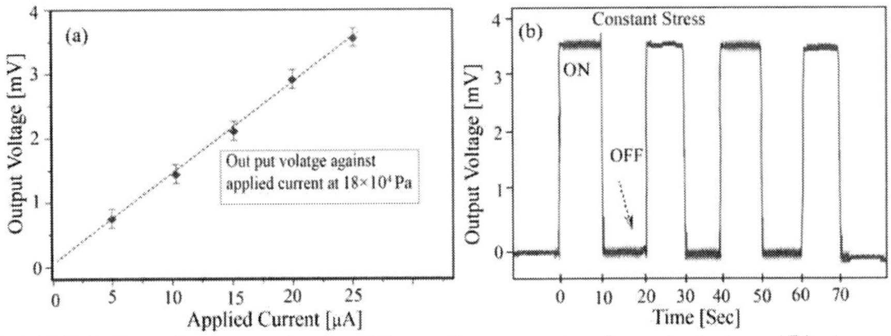

Figure 4: (a) Variation of output voltage with varying current and constant stress, (b) step response of the device showing repeatable behaviour.

Acknowledgments

This work was performed in part at the Queensland node of the Australian National Fabrication Facility, a company established under the National Collaborative Research Infrastructure Strategy to provide nano and micro-fabrication facilities for Australia's researchers. This work has been partially supported by the Griffith University's New Researcher Grants.

References

[1]. Cai X, Peng M, Yu X, Fu Y and Zou D 2014 *J. Mater. Chem. C* **2**(7) 1184–1200.
[2]. Nie Z, Deiss F, Lui X, Akbulut O and Whitesides G M 2010 *Lab Chip* **10** 3163–3169.
[3]. Bracher P J, Gupta M and Whitesides G M 2010 *J. Mater.Chem.* **20** 5117–5122.
[4]. Mahadeva S K, Walus K and Stoeber B 2015 *ACS Appl. Mater. Interfaces* **7** 8345-8362.
[5]. Akter T, Joseph J and Kim W S 2012 *IEEE Electron Device Lett.* **33**(6) 902–904.
[6]. Liu X, Mwangi M, Li X, O'Nrien M and Whitesides G M 2011 *Lab Chip* **11**(13) 2189–2196.
[7]. Lin C W, Zhao Z, Kim J and Huang J 2014 *Sci. Rep.* **4** 3812.
[8]. Liao X, Liao Q, Yan X, Liang Q, Si H, Li M and Zhang Y 2015 *Adv. Funct. Mater.* **25**(16) 2395–2401.
[9]. Phan H P, Dao D V, Dinh T, Brooke H, Qamar A, Nguyen N T, Zhu Y 2015 *Proc. 28th IEEE Int. Conf. Micro Electro Mechanical Systems (MEMS)*, Estoril 825–828.
[10]. Kang T K 2014 Appl. Phys. Lett. 104 073117.
[11]. Dinh T, Phan H P, Dao D V, Woodfield P, Qamar A, Nguyen N T 2015 J. Mater. Chem. C **3** 8776-8779

[12]. Dinh T, Phan H P, Qamar A, Nguyen N T, Dao D V 2016 RSC Advances **6** (81) 77267-77274
[13]. Mian A, Suhling J C and Jaeger R 2006 IEEE Sens. J. **6** 340–356
[14]. Doelle M, Mager D, Ruther P and Paul O 2006 Sens. Actuators A **127** 261.
[15]. Qamar A, Dao D V, Han J, Phan H P, Younis A, Tanner P, Dinh T, Wang L, Dimitrijev S 2015 Journal of Materials Chemistry C **3** 12394-12398.
[16]. Qamar A, Phan H P, Han J, Tanner P, Dinh T, Wang L, Dimitrijev S, Dao D V 2015 Journal of Materials Chemistry C **3** 8804-8809.
[17]. Qamar A, Phan H P, Dao D V, Tanner P 2015 IEEE Electron Device Letters **36** 708-710.
[18]. Phan H P, Qamar A, Dao D V, Dinh T, Wang L, Han J, Tanner P, Dimitrijev S, Nguyen N-T 2015 RSC Advances **5** 56377-56381.
[19]. Kanda Y and Yasukawa A 1982 Sens. Actuators **2** 283.
[20]. Kanda Y and Yamamura K 1989 Sens. Actuators **18** 247.
[21]. Kanda Y and Migitaka M 1976 Phys. Status Solidi A **38** K41.
[22]. Ren T L, Tian H, Xie D and Yang Y 2012 Sensors 12 6685-6694.

Cantilever-enhanced photoacoustic spectroscopy applied in the research of natural and synthetic calcium phosphate

A Brangule[1,2], K A Gross[2], V Stepanova[3]

[1]Dept.of Human Physiology and Biochemistry, Riga Stradiņš University, Dzirciema16, LV-1007, Riga, Latvia
[2]Biomaterials Research Laboratory, Riga Technical University, P.Valdena 3, LV-1048, Riga, Latvia
[3] Institute of General Chemical Engineering, Riga Technical University, P.Valdena 3, LV-1048, Riga, Latvia

E-mail: agnese.brangule@rsu.lv

Abstract. This study demonstrates the significant potential of cantilever-enhanced Fourier transform infrared photoacoustic spectroscopy (FTIR PAS) principles. The improved sensitivity and reproducibility of this method presents a potent tool in the study of biomaterials. The article discusses aspects of the application of cantilever-enhanced FTIR PAS in the research of natural and biological calcium phosphate and the statistical evaluation of the FTIR PAS sampling method. The improved constructions of the FTIR PAS accessory reduce limitations of the conventional capacitive microphone and provide a sensitive tool for samples or processes unreachable by more traditional transmittance methods, or ATR sampling technique. The most common and important applications have been discussed in-depth to show the wide range of problems solved by FTIR PAS.

1. Introduction

This article discusses the aspects of the validation of the Fourier transform infrared cantilever-enhanced photoacoustic spectroscopy's method and its application in the characterization of natural and synthetic calcium phosphate. Nanosized carbonated calcium phosphate (amorphous and crystalline) play an important role in the formation of natural and synthetic biomaterials [1]. Various techniques, such as x-ray diffraction (XRD) and spectroscopic methods (IR and Raman) can be applied in the characterization of calcium phosphates [2-4]. In previous studies, the FTIR spectroscopy's method (especially the transition mode) was the most widely used method because it can display characteristic absorption peaks over a large range of crystallinity [2,5]. The main problem with samples, when using FTIR in the characterization of calcium phosphates, is that almost all solid materials are too opaque for the direct transmission mode. This problem can be solved by reducing the optical density of samples by mixing or pressing the powder with KBr, or using various sampling techniques [6,7].

This article discusses the aspects of the cantilever-enhanced FTIR PAS application in the research of natural and biological calcium phosphate. The major advantage of the photoacoustic effect is the fact that sensitivity is not dependent on the optical path length. This allows high sensitivity from short absorption path length, and a highly linear concentration response over a wide dynamic measurement range, from very low sample volumes [8,9]. In previous works in the characterization of calcium phosphates the sensor used was a pressure sensor, the condenser microphone, which mainly limits the sensitivity of the photoacoustic detection [10]. Our study is based on PAS with an interferometer cantilever detector that allows for increased sensitivity, even several orders of magnitude higher than the condenser microphone. Due to the high detector sensitivity, PAS can work on a small scale, previously reached only by IR microscopy. In the scale of photoacoustic spectroscopy, the samples are considered remarkably small, and are hence called micro samples [11]. Previous studies reported that PAS with a cantilever detector is a valuable tool for studying samples of various morphologies due to the ease of sample preparation and depth profiling capabilities [9]. This study is divided into two parts: a validation of the FTIR PAS method and a comparison with FTIR transmission and applications, focusing on the advantages of PAS and the limitations of FTIR transmission.

2. Materials and Methods

2.1. Characterization Methods

2.1.1. PAS-FTIR spectroscopy (FTIR PAS). PAS spectra were taken at 450 – 4000 cm^{-1} at a resolution of 4 cm^{-1}, and the average was made from 10 scans with Gasera PA301, with the cell being filled with helium gas (flow 0.5 l/min). A special preparation method was not required for the synthesized powders and bone powders; 0.01g of powder was placed in the PAS cell. The pressed calcium phosphate pellet was placed in the PAS cell as well.

2.1.2. FTIR Transmission (FTIR KBr). Transmission spectra were taken at 450 – 4000 cm^{-1} at a resolution of 4 cm^{-1}, and the average made from 16 scans with the Perkin Elmer Spectrum Two. Synthesized powders were pressed into KBr pellets.

2.1.3. X-ray powder diffraction (XRD). Diffraction patterns were recorded entirely for the powders used in the validation process with the Bruker D8 ADVANCE diffractometer, from 5° to 60° using Cu Kα radiation (λ = 1.54 Å generated at 40 mA and 40 kV) at a step size of 0.2°. The crystallinity and crystallite size was evaluated using Profex 3.7.0 software [12].

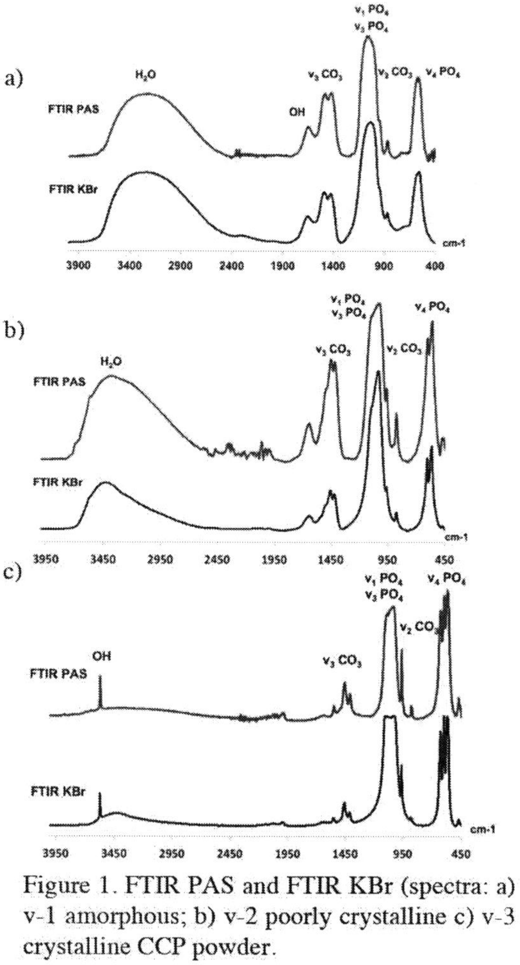

Figure 1. FTIR PAS and FTIR KBr (spectra: a) v-1 amorphous; b) v-2 poorly crystalline c) v-3 crystalline CCP powder.

2.1.4. Analysis of spectra. The FTIR spectra were viewed and smoothed with the freeware software *Specwin32*, *SpectraGryph 1.0* and *Grams/AI*. The baseline correction and curve-fitting analysis were performed using the software *MagicPlotStudent*. Deconvolution and convolution involved both Lorentzian and Gaussian curve fitting. Statistical analyses, such as the Pearson Product-Moment Correlation Coefficient (PPMCC), Cluster Analysis (CA) Chebischev Distance Matrix and the Principal Component Analysis (PCA) were performed using software Statistica 10.0.

2.2. Materials

2.2.1. Synthesis. Carbonate containing calcium phosphate (CCP) powders were obtained using following reactants: a solution containing $Ca(NO_3)_2$ and 30% ammonia mixed with a solution containing $(NH_4)_2HPO_4$ and $(NH_4)_2CO_3$ at room temperature, filtered and washed several times with deionized water containing ammonia.

2.2.2. Powders for the validation process. To adjust the particle size and composition of powders, drying was conducted in a freeze-drier at -50 °C for 72 h or in a convection oven at 200 °C. A part of the freeze-dried powder was heated at 900 °C. The powders were labelled as *v-1*, *v-2*, *v-3* according to the particle size, where v-1 represents the amorphous powder (particle size <1nm), v-2 depicts a poorly crystalline (particle size 20 nm), but v-3 is a crystalline powder (particle size 70 nm).

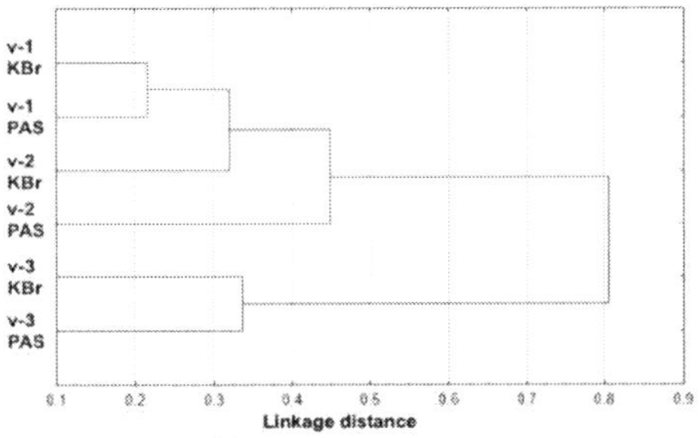

Figure 2. Hierarchical cluster method (Chebischev Distance Matrix) for FTIR PAS and FTIR KBr and amorphous, poorly crystalline and crystalline powders in 400 – 1800 cm^{-1} region.

3. Results and Discussions

3.1. Validation of FTIR PAS.

In order to validate the usage of PAS spectroscopy in the field of calcium phosphate, PAS spectra were recorded for the three most typical nanosized CCP powders: amorphous (v-1), poorly crystalline (v-2), and crystalline (v-3). To evaluate this method, FTIR PAS spectra were compared with the transmission FTIR spectra for the same CCP powders. The CCP samples of both FTIR sampling methods were examined in the wavenumber range 400 – 4000 cm^{-1} (Fig. 1). A comparison of FTIR PAS and FTIR KBr methods by qualitative features showed at well-defined peaks: 460 – 700 cm^{-1} (υ_4 PO_4, υ_L OH), 900 – 1200 cm^{-1} (υ_1, υ_3 PO_4) and 1340 – 1800 cm^{-1} ($\upsilon_3 CO_3^{2-}$ non-apatitic, type A and B) for both methods and all three powders; the presence of OH bands at 3750 cm^{-1} for both methods in the crystalline powder (v-3) and the presence of H_2O at 2500 – 3900 cm^{-1} for amorphous powders (v-1 and v-2). FTIR PAS method is more sensitive to $\upsilon_2 CO_3^{2-}$ group detection in the 860 – 890 cm^{-1} region. PAS spectra were noisy in 1950 – 2450 cm^{-1} region, but they have no effect on CCP characteristic groups in υ_2, υ_3 CO_3^{2-}, υ_4 PO_4^{3-}, υL OH/H_2O band region. A chemometric analysis of FTIR spectra was conducted in the region with an analytical significance of 400 – 1800 cm^{-1}. Characterizations with PPMC coefficients were performed at the 460 – 700 cm^{-1} (υ_4 PO_4, υ_L OH) and 1340 -1800 cm^{-1} ($\upsilon_3 CO_3^{2-}$ non-apatitic, type A and B) region. The calculated PPMC coefficients in the defined spectral regions show a strong comparability of methods (Table 1). A cross-validation with PPMCC was performed between amorphous (v-1) and crystalline (v-3) powders recorded by PAS and FTIR KBr mode (Table 1). In the cross-validation process, PPMCC shows a differentiation between amorphous and crystalline

Table 1. PPMCC of PAS FTIR vs FTIR KBr spectra

	Spectra region	460 – 700 cm^{-1} (υ_4 PO_4, υ_L OH)	1340–1800 cm^{-1} ($\upsilon_3 CO_3^{2-}$ non-apatitic, type A and B)
Validation	v-1 Amorphous	0.991	0.970
	v-2 Poorly crystalline	0.963	0.969
	v-3 Crystalline	0.979	0.986
Cross-validation	PAS (v-1; amorphous) vs FTIR KBr (v-3; crystalline)	0.730	0.867
	PAS (v-3; crystalline) vs FTIR KBr (v-1; amorphous)	0.712	0.817

powder. The Chebyschev Distance Matrix (Figure 2) confirms that both methods show a significant difference between amorphous and crystalline CCP powder, and can be used in their characterization.

The validation method, being a very important step in PAS, was applied for the process characterization. In our research PAS spectra were recorded for amorphous (v-1) powder, which was heated at different temperatures for a constant period of time. Characteristic bands were assigned to literature data [2,13]. Spectra obtained in the 460 – 700 cm^{-1} (υ_4 PO_4, υ_L OH), 1340 – 1600 cm^{-1} ($\upsilon_3 CO_3^{2-}$ non-apatitic, type A and B) and 3500 cm^{-1} – 3700 cm^{-1} (OH) region show a considerable change in peak shape and position (Fig. 3). Additional information on CCP powders can be obtained from all three regions: structure, composition, and level of crystallinity. By splitting υ_4 PO_4 peaks into a well-defined doublet, the rising of the third band 630 cm^{-1}, the rising of OH 3750 cm^{-1} bands, and changes of to the A-type and B-type carbonates can be interpreted as higher crystallinity and a transformation to hydroxyapatite structure. The results of the study obtained by FTIR PAS confirmed the usage of this method in the characterization of the process.

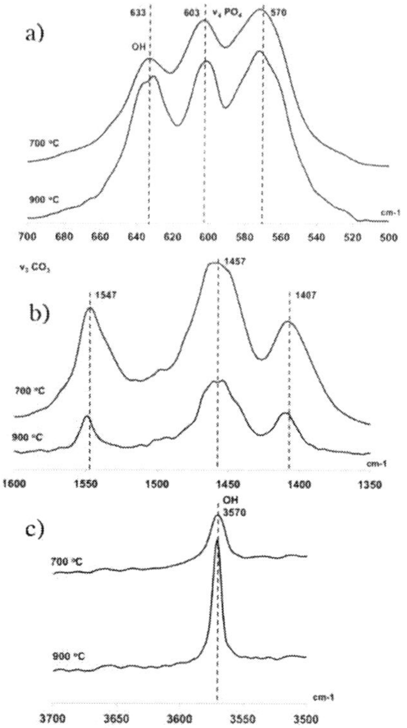

Figure 3. Effect on heating temperature on v-1 amorphous CCP powder (in the PAS spectra region: a) 460 – 700 cm^{-1}; b) 1340 – 1600 cm^{-1} ; c) 3500 cm^{-1} – 3700 cm^{-1}.

3.2. *Application of FTIR PAS.*

Three methods are presented, each one typical for a particular area of application, or of an important consideration in FT-IR PAS measurements. The methods were the following:

1. Process characterization on the surface.
2. Characterization of wet powders.
3. Characterization of microsamples.

3.2.1. *Process characterization on the biomaterial surface during the bacteria colonization process.*

Microbial analysis was chosen as a model by which we wanted to observe two processes simultaneously: a presence of functional groups of pressed amorphous CCP pellet and a presence of bacteria on its surface. Samples were prepared according to the method described in previous

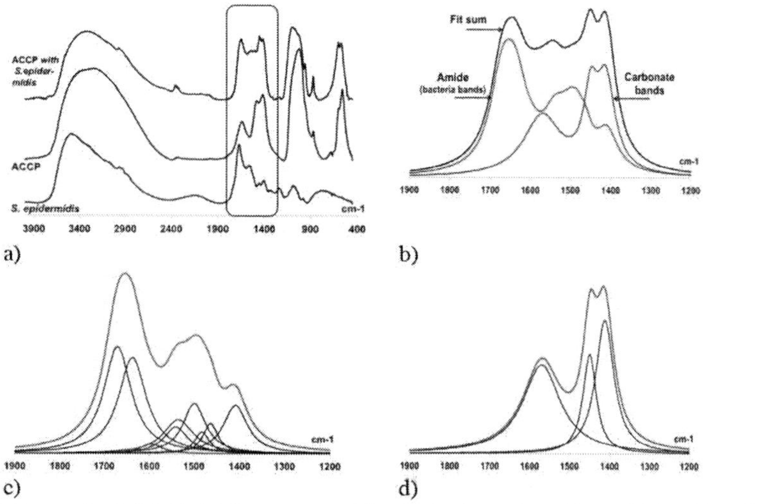

Figure 4. a) FTIR PAS spectra of bacteria *S.epidermidis*, CCP and CCP surfaceand bacteria; b) The Deconvolution of ACCP surface with S.epidermidis showing deconvoluted and convoluted c) Amide band and d) Carbonate band.

studies [14,15]. Spectra were obtained on the surface of the pressed CCP pellet before and after the colonization of the *S.epidermidis* bacteria and of the separately grown bacteria (Fig. 4a). The novelty of the application of the sensitive cantilever enhanced FTIR PAS method is the opportunity to simultaneously identify carbonate bands (1400–1540 cm^{-1}; υ_3CO_3 A, B), OH bands (1640 cm^{-1}) from CCP

Table 2. Advantages and disadvantages of cantilever PAS-FTIR

Advantages	Disadvantages
• Direct surface measurement; • No effect of peak over saturation; • Very high reproducibility; • No risk of sample contamination; • Measure possibility of measurement of samples with high humidity; • Research possibility of research of reaction transformation, crystallization by direct measurement.	• Additional cost of unite; • Limited size of samples (D ~ 1 cm); • Higher spectral noise.

surface, and amide I, II, III bands which suggested the presence of *S.epidermidis* bacteria [14,16,17]. The determination of characteristic bands is limited by overlapping amide, carbonate and OH bands in the 1200 – 1900 cm^{-1} region (Fig. 4b). The only deconvolution of spectra shows detailed information about inorganic (CO_3, OH from CCP) and organic (amides in bacteria) matter (Fig. 4c; 4d).

3.2.2. Characterization of wet powder. A spray-drying method was chosen to obtain amorphous CCP powders with a constant moisture content (9%). A wet sample analysis allows for time saving that is spent on drying (2h) and to evaluate faster the next steps in the powder processing more quickly. Other FTIR sampling methods (transmission, KBr) can't be used due to the drying effect of KBr and the fast drying and crystallization of a sample on the DRIFT sampling stick. During this experiment, FTIR PAS spectra were recorded for wet CCP with and without contamination with nitrate and ammonia ions (Fig. 5), because contamination with nitrate and ammonia ions can proceed if synthesized samples are not properly washed with distilled water after synthesis. The obtained PAS spectra show characteristic CCP bands in both poor and contaminated powder. In contaminated powder, the peaks near 3000 - 3500 cm^{-1}, 2500 cm^{-1}, 1000 - 1500 cm^{-1} suggest the presence of nitrate and ammonia ions. FTIR PAS is sensitive to nitrates and spectra can be recorded for wet CCP powders.

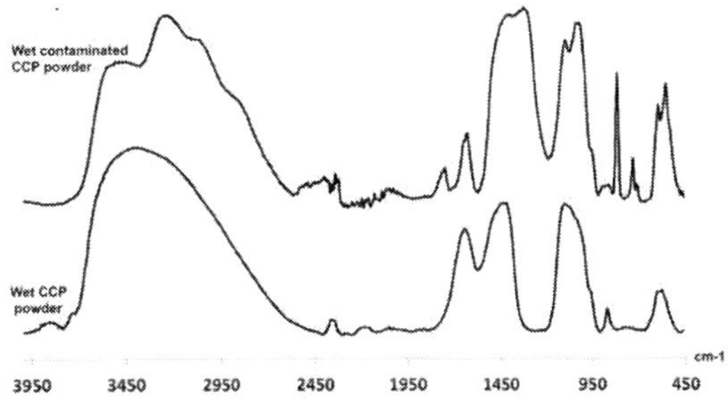

Figure 5. FTIR PAS of wet poor and wet contaminated CCP.

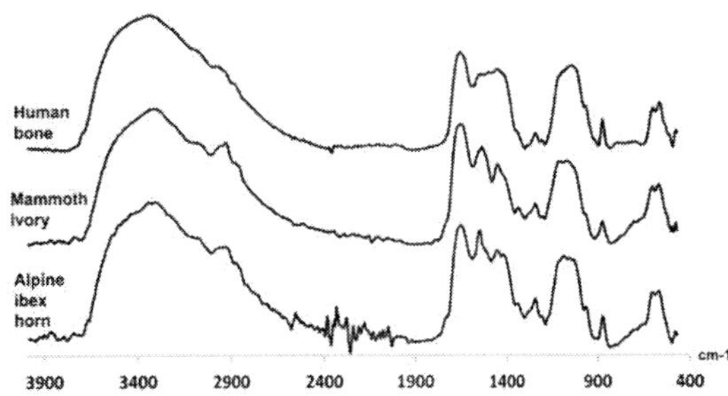

Figure 6. FTIR PAS spectra of biological samples: human bone, mammoth ivory, and Alpine ibex horn.

3.2.3. Characterization of microsamples. Artefacts and organic gem materials are stored in various collections or museums. Researchers should be very accurate during the process of studying and analysing them while being as non-destructive as possible. At the same time, they should be able to answer questions concerning their identification, particularly – what is this object made of; postulate provenance, and authenticity [18]. In the last method, we wanted to represent cantilever enhanced FTIR PAS as a sensitive, representative, and non-destructive tool with minimum sample preparation. The recorded spectra of archaeological human bone and organic gem material showed well-defined peaks: 460 – 700 cm^{-1} (υ_4 PO_4, υ_L OH), 900 – 1200 cm^{-1} (υ_1, υ_3 PO_4) and the presence of H_2O at 2500 – 3900 cm^{-1}. The 1000 – 1900 cm^{-1} region has an overlap of amide and CH_2 peaks (from organic matter), with carbonate peaks (from inorganic matter), which can be distinguished only by deconvolution [19].

4. Conclusions

This comparative study shows significant potential for cantilever-enhanced photoacoustic spectroscopy. The improved sensitivity and reproducibility of the method, gives a potent tool for the study of reaction kinetics and the transformation of compound phases. The none-destructive nature of the method provides advantages for the study of samples, where the need to conserve material or study of fixed points exists. At the same time, the method has significant limitations. Notably, sample size limitation, reduced signal/noise, and the additional cost of the sampling device. A photoacoustic sampling device gives valuable information for complicated samples, which are unreachable for ATR, DRIFT or transmittance sampling equipment.

References

[1] Legeros R Z 1994 In Brown P W, Constantz B (Eds.), Hydroxyapatite and Related Materials, Boca Raton, CRC Press, pp 3-28.
[2] Eichert D *et al* 2007 In Kendall JB (ed) Biomaterials research advaces. Nova Science, New York, pp 93-145.
[3] Gadaleta S J *et al* 1996 *Calcif Tissue Int* **58** 9–16.
[4] Ślósarczyk A *et al* 2005 *J. Mol. Struct.* **744–747** 657–61.
[5] Rey C *et al* 1991 *Calcif Tissue Int* **49** 251.
[6] Rockley M G *et al* 1980 *Spectrosc.* **34** 405-406.
[7] Colthup N B *et al* 1990 Introduction to infrared and Raman specroscopy, 3rd Edn, Academic Press.
[8] McClelland J F *et al* 2006 Photoacoustic Spectroscopy. In Handbook of Vibrational Spectroscopy.
[9] Uotila J *et al* 2008 *Appl. Spectrosc.* **62** 655-660.
[10] Rehman I *et al* 1997 *J. Mater. Sci.* **8** 1-4.
[11] Lehtinen J *et al* 2013 *Appl. Spectrosc.* **67**, 846–850.
[12] Döbelin N *et al* 2015 *J. Appl. Crystallogr.* **48**, 1573-1580.
[13] Fleet M E *et al* 2004 *Am. Mineral.* **89** 1422-1432.
[14] Brangule A *et al* 2016 *Key Eng. Mater.* **720** 125-129
[15] Slobbe L *et al* 2009 *J. Clinical Microbiol.* **47** 885–888.
[16] Kaflak A *et al* 2011 *J. Molec. Struct.* **997** 7-14.
[17] Ardeleanu M *et al* 1992 in: Springer Series in Optical Sciences **69**, Springer-Verlag Berlin, Heidelberg pp 81-84.
[18] Gu C *et al* 2013 *Spectrochim. Acta. A. Mol. Biomol. Spectrosc.* **103** 25–37.
[19] Brangule A *et al* 2015 *IOP Conf. Series: Mater. Sci. Eng.* **77**.

Investigation of porous Si formed by metal-assisted chemical etching with Au as catalyst

O. Pyatilova[1]*, S. Gavrilov[1], A. Savitskiy[1], A. Dudin[2], A. Pavlov[2], Yu.Shilyaeva [1]
[1]National Research University of Electronic Technology, Zelenograd, Moscow, Russia
[2] Institute of nanotecnology of microelectronics of the Russian academy of sciences, Moscow, Russia
*e-mail: 5ilova87@gmail.com

Abstract.
In this work, we investigated the effect of the reaction zone temperature and amount of catalyst on the reaction kinetics of Si chemical etching using Au clusters. The growth rates of the porous layer thickness (V_h) and pore diameter (V_d) was calculated. Influence of etching process parameters on the pore's sidewall angle was established.

1. Introduction

Porous Si has a wide range of potential applications in nanoscale electronics [1, 2], optoelectronics [3], thermoelectrics [4], photovoltaics [5], biosensors [6], and as electrodes in Li-ion batteries [7]. Metal-assisted chemical etching (MACE) is a novel method that can be successfully used to form nanoscale structures in Si [8-10]. This approach is based on the use of metal films or clusters loaded on Si wafers as catalysts for etching of Si in HF/H_2O_2 aqueous solutions without external electrical power. Etching is initiated by the metal, thus the kinetics of process and the geometry of the porous layer are affected by the type and structure of metal catalyst. MACE process can be described by means of the galvanic corrosion in which Si is anode and Au is cathode [11, 12]. Therefore, the cathode area influences the anodic dissolution rate of silicon. On the other hand, MACE process can be described in terms of band theory, since there is a metal-semiconductor interface [13]. In this case, as is known, the temperature influences the concentration of charge carriers and injection of holes from a metal and this requires detailed study for MACE process. In this work, we investigate the effect of the reaction zone temperature and amount of catalyst on the reaction kinetics of Si chemical etching using Au clusters.

2. Experiment

Wafer of *p*-type Si ((100), 0.01 Ω·cm) was used as substrate. The single-crystalline wafers were rinsed with piranha etch (mixture of H_2SO_4 (98%) and H_2O_2 (30%) (1:2 in volume)) at 130°C during 10 min prior to etching. Then wafers were rinsed with deionized water and dried in the isopropyl alcohol vapor flow. After that, Si wafers were rinsed in HF-H_2O solution (1:5 in volumes) in order to remove native SiO_2. Au clusters were deposited on Si wafer using the vacuum thermal evaporation method. In order to prepare the samples, the two weights of Au were subsequently evaporated on 100 mm-Si wafer: 15 mg and 12 mg. Test samples covered with Au clusters were placed into solution containing HF (40%)/H_2O_2(30%)/H_2O (volume ratio 17/3/51) at 25°C, 40°C, 50°C and 70°C for 10-50 min [14]. Then samples were rinsed in ethanol and air-dried. Morphology of the sample surface was investigated by scanning electron microscopy (SEM) using Helios NanoLab 650. Surface composition of samples was analyzed by energy-dispersive X-ray spectroscopy (EDX). In order to estimate geometrical parameters of the metal catalyst, the cluster diameter distribution histograms were obtained by computer processing of the SEM images of the Au coated wafer surface using Axio Vision software.

3. Result

Figure 1 shows the SEM images of the Si surface covered with Au cluster array after the consecutive evaporation of 15 (Figure 1a) and 12 mg weight portions of gold (Figure 1b). The insets show the gold cluster diameter distribution histograms per 1 μm^2.

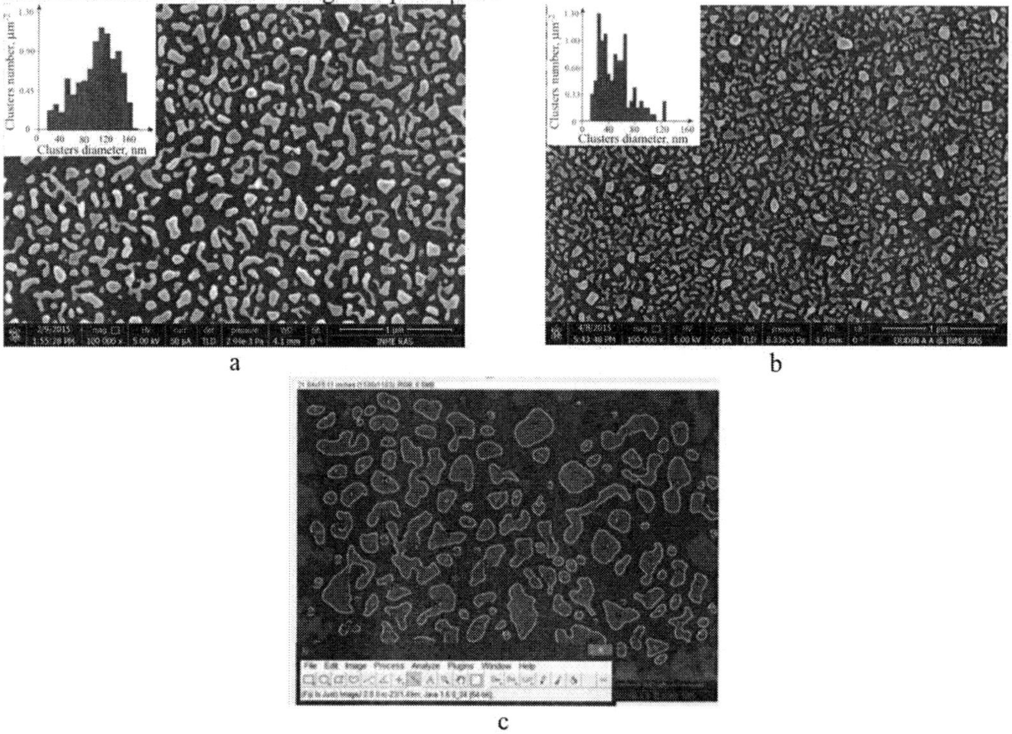

Figure 1. SEM images of sample surface after the consecutive evaporation of 15 mg (a) and 12 mg (b) of gold and corresponded cluster diameter distribution histograms; Axio Vision dialog box while processing of SEM images (c)

As it follows from the results of the microscopy, the subsequent evaporation of the 15 mg and 12 mg of gold leads to the formation of clusters with prevailing diameter of 100-120 nm; 30 nm and 50 nm, respectively. The SEM cross-sectional views of por-Si structures obtained after etching of Si/Au samples at different temperatures are shown in Figures 2-5. Evaporation weight portion of gold is 15 mg.

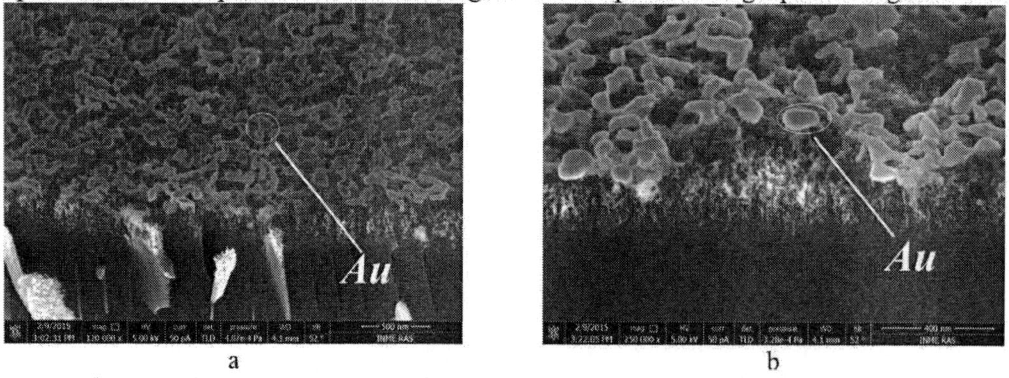

Figure 2. SEM images of por-Si formed at 25°C during 10 min (a) and 50 min (b).

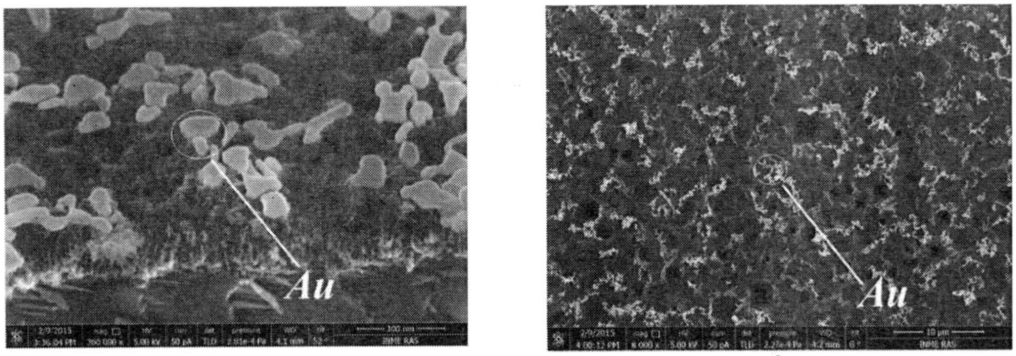

a b

Figure 3. SEM images of por-Si formed at 40°C during 10 min (a) and 50 min (b).

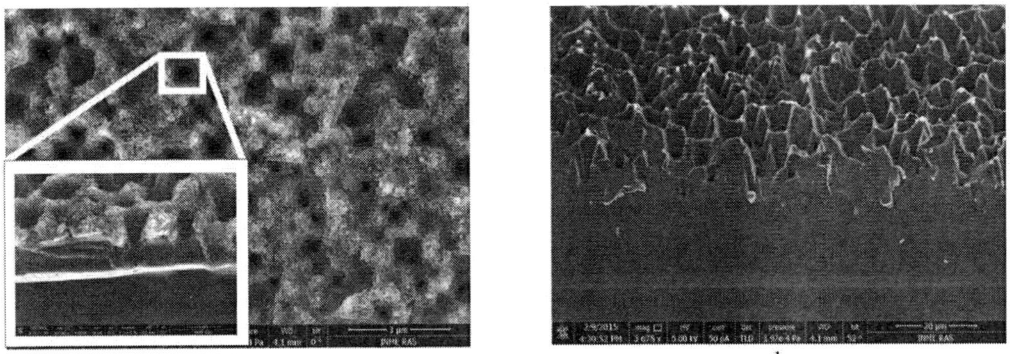

a b

Figure 4. SEM images of por-Si formed at 50°C during 10 min (a) and 50 min (b).

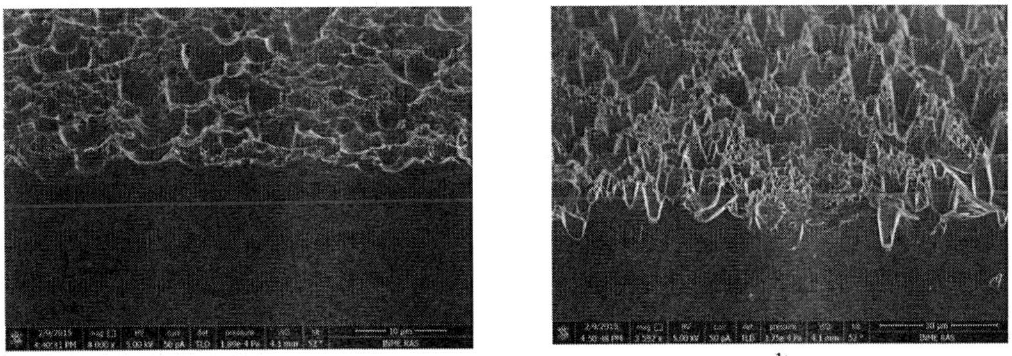

a b

Figure 5. SEM images of por-Si formed at 70°C during 10 min (a) and 50 min (b).

The etching process at 25 °C leads to formation of the pores oriented along the crystallographic direction [100]. Additionally, the small side branches of pores are also formed. The geometrical parameters (thickness of porous layer and pores diameter) of the por-Si were determined on the basis of the results of computer processing of SEM images. Figure 6 illustrates that the thickness of the porous layer depends on the etching duration. In case of etching process at 25 °C, the thickness of the porous layer is 250 nm, 370 nm and 580 nm after 10 min, 30 min and 50 min etching, respectively. The growth rate of the porous layer increases threefold with increasing reaction zone temperature up to 40 °C. 30-min etching leads to formation of square pores with a side of 0.25-1.63 μm. Increasing the solution temperature up to 50 °C and 70 °C promotes the formation of porous layer with the thickness of 2, 6 and 10 μm after 10 min, 30 min and 50 min etching, respectively.

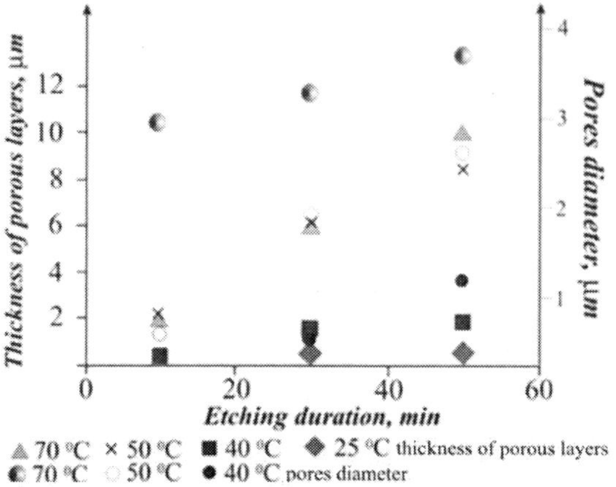

Figure 6. Dependence of the thickness of porous layer and pore diameter on etching duration at different temperatures (from 25 °C to 70 °C).

Thus, the porous silicon layers were formed with a wide range of geometrical parameters: thickness from 0.24 μm to 10 μm; pore diameter from tens of nanometers to 3.5 μm. The thickness of the porous layer and the pore diameter increase with increasing temperature of the solution. This is due to the following factors:

- increasing charge carrier concentration on semiconductor-electrolyte interface;
- increasing diffusion coefficient of reagents and products in the solution;
- convective mass transfer occurring because of the temperature gradient;
- increasing hole current due to intensive injection h+ from metal by reducing the width of the potential barrier Au/Si.

In this work we calculated the etching rate of monocrystalline silicon at various temperatures. The etching rate of silicon means two parameters: the growth rate of the porous layer thickness (V_h) and the growth rate of the pore diameter (V_d). If reaction rate constant of silicon etching is known, one can then predict the geometry of the porous layer.

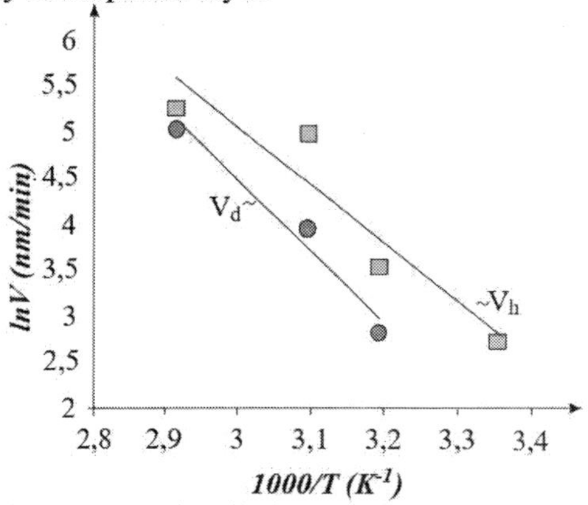

Figure 7. Experimental plots for thickness of porous layer and pore diameter.

Accordingly, an activation energy (E_a) and pre-exponential factor (A) have been calculated using the Arrhenius equation [10]. The growth rate of the porous layer thickness is described by the following equation: $V_h=2{,}6\cdot10^{10}\cdot\exp(-0{,}54/kT)$, where V_h - reaction rate coefficient, $E_{ah}=0{,}54$ eV - the activation energy of growth, A_1=2,6·10^{10} nm/min is pre-exponential factor at E_{ah}. The growth rate of the pore diameter is described by the following equation: V_d=1,01·10^{12}·exp (-0,66/kT), where E_{ad}=0,66 eV, A_2 =1,01·10^{12} nm/min at E_{ad}.

Additionally, pore's sidewall angles have been measured for the samples of por-Si formed at 50°C and 70°C (corresponded SEM images are shown in Figures 4, 5). Figure 8a illustrates the evolution of the geometry of por-Si layers with increasing the etch temperature and the duration of etching.

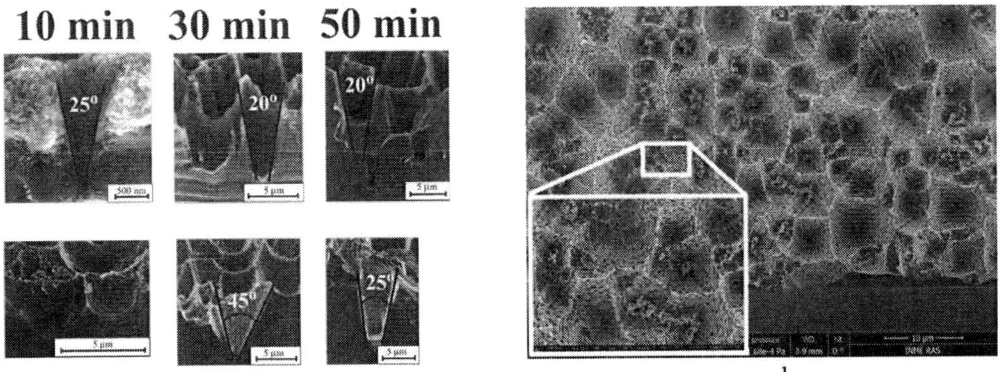

Figure 8. The evolution of the geometry of por-Si layers (a), SEM top view of por-Si (b)

It is established that the pore's sidewall angle decreases with increasing duration of etching. This is due to the difference in the dissolution rates at the pore bottom and top because of the different concentrations of Au. Figure 8 b presents SEM top view of por-Si after 30-min etching process at 70 °C. It should be noted that the pores are formed with a bottom parallel to the crystallographic direction [100] for all samples after 50-min etching at 50 °C and 70 °C. It is established that Au clusters are present at the bottom of the pores. Since the etching process is Au-assisted, the concentration of Au affects the dissolution rate [15]. Therefore, the dissolution rate at the bottom of the pores is higher than at the surface of sample and V_h>V_d.

As the Au-assisted chemical etching in itself is local electrochemical etching, the cathode area contributes to the value of Si dissolution rate [16]. The thickness of *por*-Si layer depending on the duration of etching at different temperatures (from 25 °C to 50 °C) is shown in Figure 9. Evaporation weight portion of Au is 12 mg.

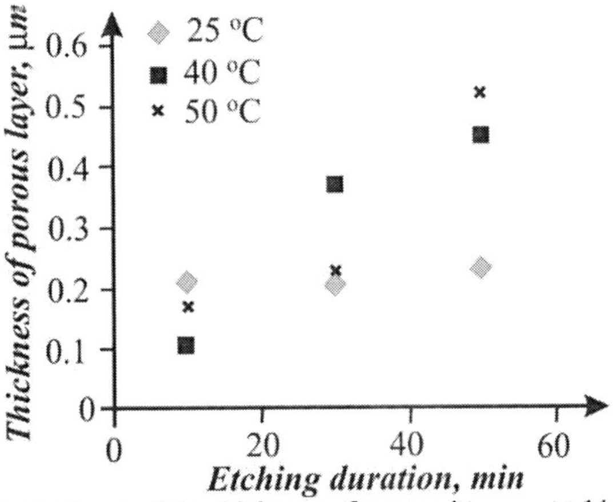

Figure 9. Dependence of the thickness of porous layer on etching duration.

It is shown, that the etching process at 25-50 °C leads to formation of the structure with pores oriented along the crystallographic direction [100]. The thickness of the porous layer depends on the etching time (10, 30 and 50 min) and is 197 nm, 200 nm and 234 nm for 25 °C; 106 nm, 369 nm and 447 nm for 40 °C; 177 nm, 230 nm and 529 nm for 50 °C, respectively. It was noted that the macroporous silicon began to form only after etching at 70 °C.

Thus, the growth rate of the porous Si layer is significantly affected by the evaporated weight portion of Au. This may be related to the dependence of cluster diameter on the amount of evaporated metal.

Conclusion

The reaction zone temperature and amount of catalyst on the reaction kinetics of Au-assisted chemical etching was investigated. The porous silicon layers were formed with a wide range of geometrical parameters depends on temperature: thickness from 0.24 μm to 10 μm; pore diameter from tens of nanometers to 3.5 μm. The growth rate of the porous layer thickness is described by the equation: $V_h=2,6\cdot10^{10}\cdot\exp(-0,54/kT)$. The growth rate of the pore diameter is described by the equation: $V_d=1,01\cdot10^{12}\cdot\exp(-0,66/kT)$. It is established that the pore's sidewall angle decreases with increasing duration of etching. This is due to the difference in the dissolution rates at the pore bottom and top because of the different concentrations of Au. Thus, the growth rate of the porous Si layer is significantly affected by the evaporated weight portion of Au.

Acknowledgment

The reported study was funded by RFBR according to the research project No 16-33-00712 mol-a.

References

[1] V. Schmidt, H. Riel, S. Senz, S. Karg, W. Riess, U. Gösele, Small **2**, 85 (2006).
[2] J. Goldberger, A.I. Hochbaum, R. Fan, P. Yang, Nano Lett. **6**, 973 (2006).
[3] Y. Kanemitsu, Phys. Rep. **263**, 1 (1995).
[4] A.I. Hochbaum, R. Chen, R.D. Delgado, W. Liang, E.C. Garnett, M. Najarian, A. Majumdar, P. Yang, Nature **451**, 163 (2008).
[5] B. Tian, X. Zheng, T.J. Kempa, Y. Fang, N. Yu, G Yu, J. Huang, C.M. Lieber, Nature **449**, 885 (2007).
[6] Y. Cui, Q. Wei, H. Park, C.M. Lieber, Science **293**, 1289 (2001).
[7] C.K. Chan, H. Peng, G. Liu, K. McIlwrath, X.F. Zhang, R.A. Huggins, Y. Cui, Nat Nano **3**, 31 (2008).
[8] Y. Qu, L. Liao, Yu. Li, H. Zhang, Yu Huang, X. Duan, Nano Lett., **9** 4539 (2009).
[9] Ch.-L. Lee, K. Tsujino, Yu. Kanda, Sh. Ikeda, M. Matsumura, J. Mater. Chem., **18** 1015 (2008).
[10] Z. Huang, N. Geyer, P. Werner, J. De Boor, U. Gösele, Advanced Mater., **23** 285 (2011).
[11] C.M.A. Ashruf, P.J. French, P.M.M.C. Bressers, P.M. Sarro, J.J. Kelly, Sensors Actuators A: Physical, **66** 284 (1998).
[12] K.W. Kolasinski, Nanoscale Res. Lett., **9** 432 (2014).
[13] M. J. Sailor Porous Silicon in Practice: Preparation, Characterization and Applications (Wiley-VCH Verlag & Co. KGaA, Weinheim, 2012), p. 249.
[14] D. Wang, R. Ji, S. Du, A. Albrech and P. Schaaf, Nanoscale Research Letters, **8** 42 (2013).
[15] O.V. Pyatilova, S.A. Gavrilov, A.A. Dronov, Ya.S. Grishina, A.N. Belov. Sol. St. Phenomena, **213** 103 (2014).
[16] Q. L. Chang, H. Cheng, W. K. Choi, and Carl V. Thompson, J. Phys. Chem. C, 117 20802 (2013).

Platinum-free catalysts for low temperature fuel cells

Tatiana Lastovina[1], Julia Pimonova[1], Andriy Budnyk[1]

[1]International Research Center "Smart materials", Southern Federal University, 5 Zorge Street, Rostov-on-Don, Russia

Corresponding author: e-mail: lastovina@sfedu.ru

Abstract. In this work, we have successfully prepared Zn/Co-N/C and Zn/Co-Fe/N/C composites, both derived from single zeolitic imidazolate framework (ZIF) precursor Zn/Co-ZIF containing equivalent quantities of Zn and Co metal sites. The composites were formed by pyrolysis of the precursor at 700 °C in inert gas atmosphere as such and after mixing it with Fe(II) salt and 1,10-phenontraline in ethanol. Catalytic tests for oxygen reduction reaction (ORR) in electrochemical cell demonstrated promising results allowing us to consider these composites as potential Pt-free catalysts for low temperature fuel cells.

1. Introduction

Due to the limitedness of natural resources like gas and oil there is a tendency in modern world to replace the conventional energy sources with the alternative ones. Huge markets like car-building industry demand the new feasible technological solutions based on power production without burning hydrocarbons. One of the emerging technologies is based on production of electricity on board of a vehicle by electrochemical fuel cells. These low-temperature fuel cells offer a direct conversion between chemical and electrical energy, so energy losses are minimized [1].

Nowadays, the air-hydrogen fuel cells are in focus of research efforts around the globe, because they produce water as the main product of the reaction with a few byproducts. The efficiency is limited by the rate of the ORR [2]. In such cells the commercially available cathodic catalysts usually contain platinum and its alloys (e.g. Pt/C, Pt_3Co/C). Unfortunately, they are too expensive for common use, furthermore, they can degrade over time. The challenging solution for this problem is the development of efficient platinum-free electrocatalysts [3].

Different compounds have been employed as alternative catalysts for the ORR. Recently, metal-organic frameworks (MOFs) and their derivatives are emerged among the most promising ones. The advantages of MOFs include good thermal stability and endless variety of topologies. However, for certain cases it is hard to obtain chemically stable MOFs [4]. The MOFs are also known to exhibit catalytic activity in the ORR [5], but more importantly, they can serve as precursors for fuel cell catalysts. For instance, MOF HKUST-1 shows electrocatalytic activity in neutral media [6] and it was used as a precursor to produce Cu@N/C catalyst [7]. More generally, such catalytic system can be marked as M-N/C, where M is an atom of transition metal (Cu, Fe, Co) and N/C stands for nitrogen-doped porous carbon. Some of M-N/C catalysts reportedly demonstrate characteristics approaching the Pt-based catalyst performance [8]. The chemical stability of such materials remains the question of high importance. For example, Fe-N/C type catalyst showed high electrochemical activity and stability in alkaline, acidic [9] and neutral media [10].

IOP Conf. Series: Journal of Physics: Conf. Series **829** (2017) 012007 doi:10.1088/1742-6596/829/1/012007

Currently, the production of the M-N/C materials is based on the mixing of carbon sources (MOFs), metal-containing (often Fe(II)) compounds and nitrogen sources followed by pyrolysis of the mixture. Further enrichment can be done if chosen MOF already contains more than one metal. ZIF family of MOFs includes ZIF-8 containing Zn, and ZIF-67 containing Co. Considering imidazolic linker as a source of both nitrogen and carbon we may count on this MOF as an ideal pyrolysis precursor. Although, a lot of work has been done on designation the structure of the catalytic sites of the Pt-free electrocatalysts [12-13], but the complexity of the matter keeps question open.

It should be also mentioned, that pyrolysis temperature influences the state of final product. According to reported in the literature, the most effective one for Fe-N-C materials is 700 °C [8]. Another crucial characteristic is the porosity of the product. The samples with the higher porosity tend to demonstrate higher electrochemical activity [13-14]. Assuming that active site is a five-coordinated iron atom, micropores are the only place, where they can be located [15].

In this communication we describe some results on preparation and electrocatalytic activity tests of the composites containing transition metals embedded into N-doped porous carbon support.

2.1 Synthesis of the samples

2.1.1 Preparation of MOF

At the first stage MOF with both Co, Zn atoms and 2-methylimidazole (mIm) as linker was synthesized. For this purpose solution (total volume 18 ml) of dimethylformamide (DMF) containing 0.273 g of $Co(NO_3)_2 \cdot 6H_2O$, 0.308 g of mIm, 0,279 g of $Zn(NO_3)_2 \cdot 6H_2O$ was prepared. Then the resulting solution was heated in microwave oven (MARS 6, CEM, USA) for 120 min at 140 °C. Obtained powder was collected by centrifugation, washed several times with methanol, kept in methanol for 48 h and dried at 60 °C.

2.1.2 Carbonization of MOF

At the second stage prepared MOF was mixed with 0.0207 g of 1,10-phenontraline, 0.0012 g of Fe(II) acetate and ethanol. After sonication powder was filtered, washed, dried and annealed under Ar flow for 3 h at 700 °C with the heating rate of 10° C per minute.

As-prepared MOF without addition of iron salt and 1,10-phenontraline was also annealed at the same conditions.

Pure MOF will be further referred as Zn/Co-ZIF, MOF after heat treatment as Zn/Co-N/C and those obtained after heat treatment with addition of iron salt and 1,10-phenontraline as Zn/Co-Fe/N/C.

2.2 Characterization

X-ray diffraction (XRD) of the MOF was detected using Bruker D2 Phaser diffractometer (Cu k_α, λ= 0,15406nm).The Brunauer-Emmett-Teller (BET) specific surface area was determined by nitrogen adsorption-desorption at 77 K on an ASAP 2020 analyzer. Prior to the measurements, the samples were degassed at 140 °C for 12 h. Transmission electron microscopy (TEM) images were acquired on a Tecnai G2 Spirit BioTWIN microscope operated at an accelerating voltage of 120 kV.

The electrochemical measurements were performed in a typical three-electrode cell in 0.1 M KOH or 0.1 M H_2SO_4 solution with a Pt electrode and Ag/AgCl (sat.KCl solution) as the counter electrode and reference electrode, respectively. The electrochemical properties of all materials were investigated by cyclic voltammetry (CV) and liner sweep voltammetry (LSV). A rotating disk electrode was used as the substrate for the working electrode. For the preparation porous carbons-based catalytic inks 7 mg of the sample was dispersed in the solution containing 900 μl of isopropanol and 100 μl of 0.5 % Nafion solution. After sonication and stirring 28 μl of the catalyst ink was dripped onto the glassy carbon surface. In the case of pure MOF testing 3 mg of the sample was mixed 3 mg of carbon (Vulcan XC-72). The next procedure for catalytic inks preparation is the same.

Before investigation of the samples catalytic activity in ORR all materials were cycled in Ar-saturated electrolytes without rotation with a sweep rate of 50 mV s^{-1}. LSV curves were recorded at different speed rates (400, 625, 900, 1000, 1225, 1600, 1935 rpm). All the potentials reported in this work are vs. the standard hydrogen electrode (SHE).

3. Results

The XRD spectrum (Figure 1) of pure dual metal (Zn and Co) ZIF-type material (Zn/Co-ZIF) exhibits sharp peaks and are in good agreement with the literature data [16]. The diffraction peaks at 2θ = 7.4 °, 10.4 °, 12.8 °, 14.8 °, 16.5 °, 18.1 °, 22.2 °, 24.6 °, 25.7 °, 26.8 °, 29.8 ° can be indexed to the (011), (002), (112), (022), (013), (222), (114), (233), (224), (134) and (044), planes of Zn/Co-ZIF material.

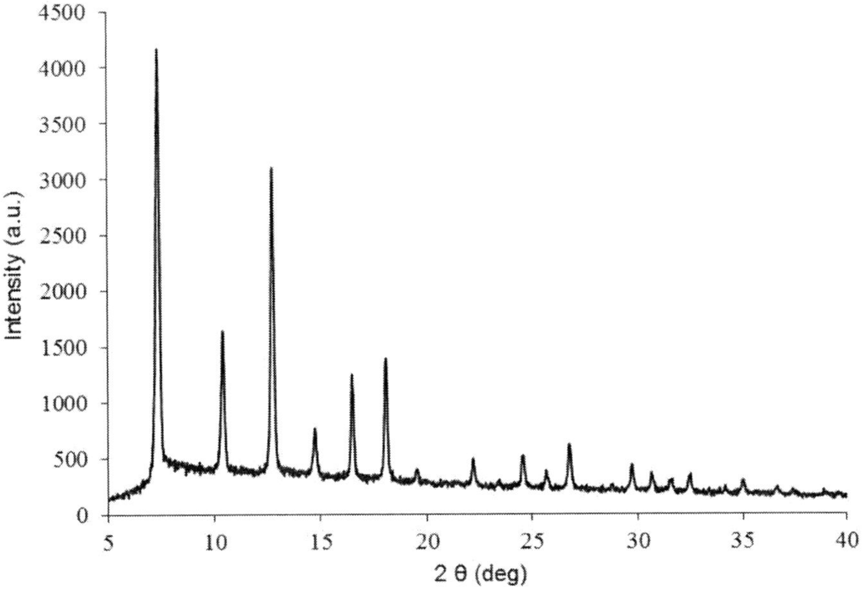

Figure 1. XRD spectrum of Zn/Co-ZIF

The surface area values for the catalyst before and after annealing were found from nitrogen physisorption at 77 K isotherms (composed of adsorption and desorption branches), presented in Figure 2. Both isotherms exhibit profiles attributed to Type I (microporous solids) according to IUPAC classification. Slight hysteresis of H4 type (IUPAC) is also observed, which in combination with given type of isotherm is known to be indicative for microporosity. A tremendous difference in gas uptake is clearly seen from the significant separation of two isotherms (~ 250 cm^2/g). The calculated values of surface areas are 1168 m^2/g (BET model) and 1406 m^2/g (Langmuir model) Zn/Co-ZIF sample, and 233 m^2/g (BET) and 284 m^2/g (Langmuir) for Zn/Co-Fe/N/C one. The reduction of the surface area due to annealing of a starting sample (which is a MOF, who typically possesses high surface area) is reasonable expected, and a decrement factor in this case is 5 times.

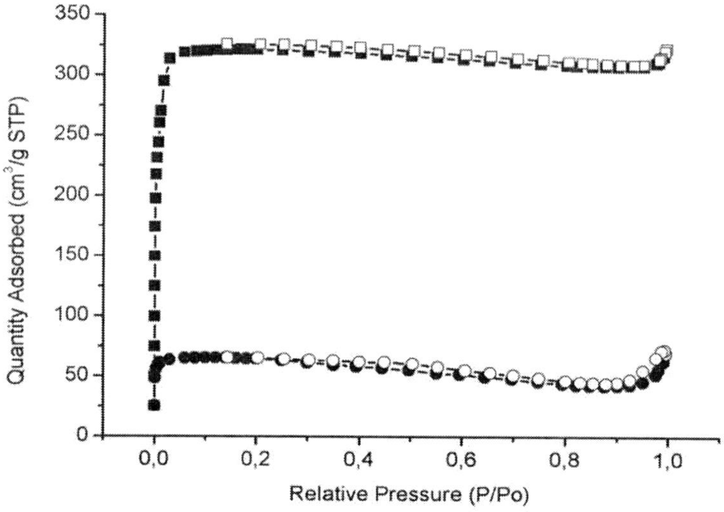

Figure 2. N_2 physisorption isotherms for Zn/Co-N/C (squares) and Zn/Co-Fe/N/C (circles) samples. The adsorption branch is shown with filled symbols, while the desorption one □ with open symbols.

The morphology of Zn/Co-ZIF and Zn/Co-Fe/N/C samples was characterized by TEM (Figure 3). It can be seen that as-prepared MOF (Figure 3A) material consisted of crystals of rhombic dodecahedral shape. In the case of Zn/Co-Fe/N/C sample, it was clearly shown that well-separated metals nanoparticles of around 7.5 nm in size are well dispersed on the carbon carrier (Figure 3B).

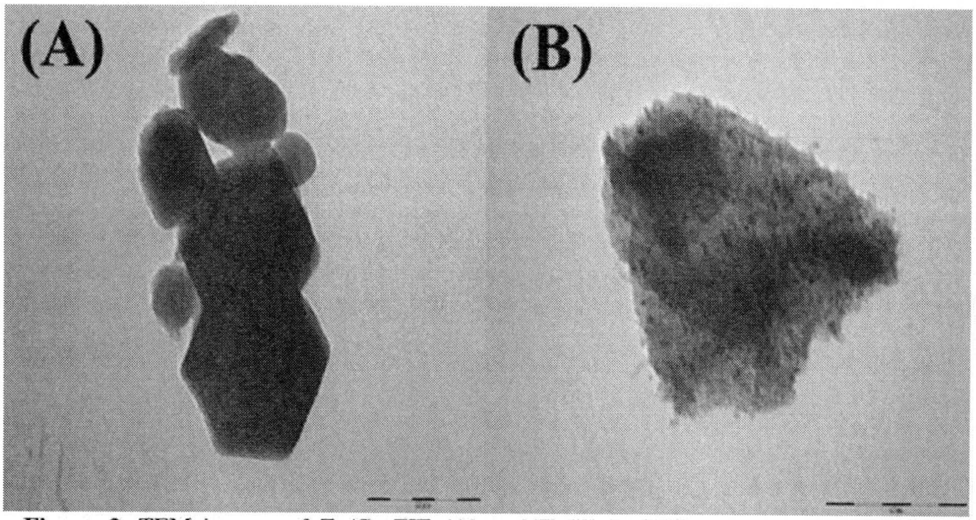

Figure 3. TEM images of Zn/Co-ZIF (A) and Zn/Co-Fe/N/C materials.The scale bar corresponds to 200 nm.

The CV curves for all materials are presented at Figure 4. It shows that electrocatalytic activity of Zn/Co-N/C material is much poorer than those of the Zn/Co-Fe/N/C catalyst in both electrolytes, especially in 0.1 M H_2SO_4 solution.Voltammetry measurements indicate that the catalytic activity of the porous carbons mainly stems from the Fe-doping. In the case of Zn/Co-Fe/N/C sample a big peak in cathodic area appears at ~0.5 V is assosiated with Fe presence on the surface of material. Also it can evidence that Fe doping can act as a mediator in the electron-transfer process for ORR [17].

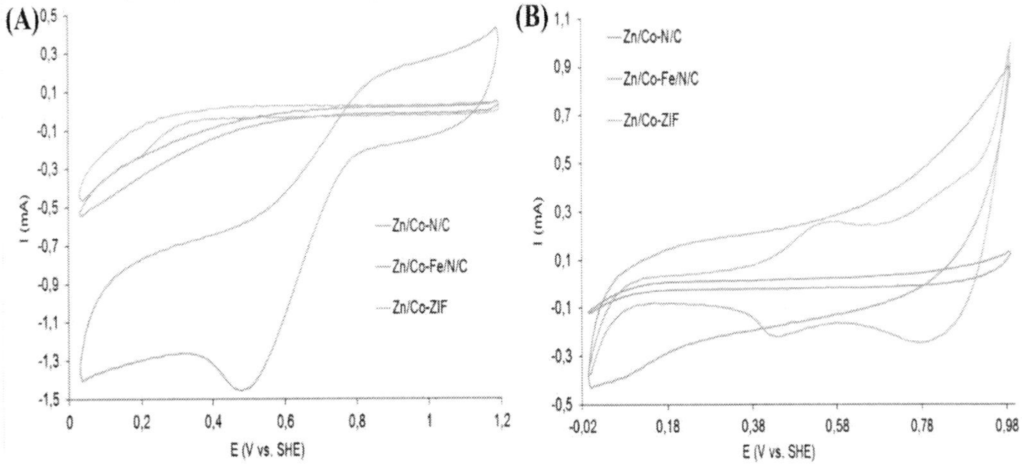

Figure 4. CVs of different samples in O_2-saturated 0.1 M H_2SO_4 (A) and 0.1 M KOH solutions at 1600 rpm at room temperature. Scan rate - 50 mV s^{-1}.

Conclusions

In summary, we report a step-by-step preparation of the electrocatalytic active Fe/N- and N-doped porous carbons. The catalytic activity of such materials is significant higher at acidic conditions. Cyclic voltammetry for ORR in three-electrode cell demonstrated promising results allowing us to consider these composites as potential Pt-free catalysts for low temperature fuel cells.

Acknowledgments

This work was supported by Russian Foundation for Basic Research (grant № 16-33-00854).

References

[1] Korovin N V 2008 *Soros Educational Journal* **10** 55
[2] Becknell N et al 2015 *J Am Chem Soc* **137** 15817
[3] Serova A et al 2015 *Nano Energy* **16** 293
[4] Kitagawa S, Kitaura R, Noro S. 2004 *Angew. Chem. Int. Ed.* **43** 2334
[5] Yuqian Ren, Guo Hui Chia, Zhiqiang Gao 2013 *Nano Today* **8** 577
[6] Mao J, Yang L, Yu P, Wei X, Mao L 2012 *Electrochemistry Communications* **19** 29
[7] Du C, Gao X, Chen W 2016 *Chinese Journal of Catalysis* **37** 1049
[8] Xiao H, Shao Z-G, Zhang G, Gao Y, Lu W, Yi B 2013 *Carbon* **57** 443
[9] Lin L, Zhu Q, and Xu An-Wu 2014 *J. Am. Chem. Soc* **136** 11027
[10] Zhi-Yu Yang, Yu-Xia Zhang, Lin Jing, Yu-Fei Zhao, Yi-Ming Yan and Ke-Ning Sun 2014 *J. Mater. Chem. A.* **2** 2623
[11] Zhu H, Zhang S, Yu-Xi Huang, Wu L, and Sun S 2013 *Nano Lett* **13** 2947
[12] Zitolo A, Goellner V, Armel V, Sougrati M, Mineva T, Stievano L, Fonda E, Jaouen F 2015 *Nature Materials* **14** 937
[13] Jing Sun, Ya-Hui Fangab, Zhi-Pan Liu 2014 *Phys. Chem. Chem. Phys* **16** 13733
[14] Li Zhang, Han Wang, Hong Wang, Xiao-Peng Liu, Liang Su, Liang Yu, and Ping Chen 2016 *J. Electrochem. Soc.* **163(9)** H738
[15] Jaouen F, Charreteur F, and Dodelet J P 2006 *J. Electrochem. Soc.* **153(4)** A689

[16] Zhou K, Mousavi B, Luo Z, Phatanasri S, Chaemchuen S and Verpoort F 2016 *J. Mater. Chem. A* DOI: 10.1039/C6TA07860E
[17] Yang M, Chen H, Yang D, Gao Y and Li H 2016 *RSC Adv.* **6** 97259

IOP Conf. Series: Journal of Physics: Conf. Series **829** (2017) 012008 doi:10.1088/1742-6596/829/1/012008

Visual analytics in investigation of chirality-dependent thermal properties of carbon nanotubes

Vadim A. Shakhnov[1], L.A. Zinchenko[1], V. V. Makarchuk[1], E.V. Rezchikova[1], V. V. Kazakov[1]
[1] Bauman Moscow State Technical University, Moscow, Russia

shakhnov@mail.ru

Abstract. In the paper, we present our approach to visual analytics application to investigation of thermal properties of carbon nanotubes. A thermal properties carbon nanotube framework is developed to investigate remarkable features of carbon nanotubes. Visual analytics is useful for comparison of their thermal properties and the corresponding analysis. The preliminary results are reported.

1. Introduction

Currently, many different mathematical models of nano-objects are used for calculation of their properties [1-5]. Using a mathematical model, a designer obtains a big numerical data file, that has to be analyzed. However, it is difficult to choose the best design solution from the two objects that are invisible. In our approach, data and designed objects are visualized [6]. A designer can analyze the obtained numerical and visual information simultaneously.

In the paper, we present an application of our approach to visual analytics support for research of carbon nanotube variation and its influence on thermal conduction phenomena. The outstanding thermal conductivity of carbon nanotubes attracts designers [1, 2, 7]. Carbon nanotubes FETs are promising candidates for the post silicon era. However, the fabrication of carbon nanotubes with the predefined performance is a big challenge [8]. Carbon nanotube specific variations, including nanotube diameter and chirality variations, influence the thermal conductivity of carbon nanotube [9]. A designer has to compare thermal properties of carbon nanotubes with different chiral indices. However, carbon nanotubes are invisible for a human eye. Therefore, special efforts are required to compare alternative design solutions.

We illustrate our approach for investigation of thermal properties of single-walled carbon nanotubes. In the paper, we proposed a novel approach based on visual analytics. The approach supports a nanotube devices design process and simplifies a design solution choice.

In the paper, we focus on design solution with the predefined diameter. First, we find all possible chiral indices for the given diameter and then visualize all possible design solutions with their thermal properties. The corresponding statistical data is reported as well.

The rest of the paper is structured as follows. The next section reviews the related works in the field of carbon nanotubes thermal conductivity. Section 3 presents our approach for visual analytics support in research of chirality-dependent thermal properties of carbon nanotubes. We use single-walled CNTs as our test case. Finally, conclusions are derived in Section 4.

2. Mathematical model of thermal conductivity of carbon nanotubes

The carbon film (graphene) and nanotubes are allotropes of carbon with unique transport properties [3, 4]. Nanotubes come in a variety of lengths L and diameters D. The distance between the adjacent carbon atoms in carbon nanotubes is equal to d_0 = 0.142 nm. Therefore, the atomic structure of a single-walled CNT is well described by a pair of indices (chiral indices), n and m. The diameter is calculated as follows:

$$D = \overline{m^2 + n^2 + m * n}\ \frac{\overline{3} * d_0}{\pi} \qquad (1)$$

The CNT thermal conductivity is dependent on the temperature, the nanotube length L and the average phonon mean free path l_0 [1, 2]. If $l_0 > L$ the ballistic conduction mechanism is dominant. A quantum of thermal conductance is given as follow:

$$G_{th} = \frac{\pi^2 k_B^2 T}{3h} = 9{,}46x10^{-13} \frac{W}{K^2} T \qquad (2)$$

where k_B is the Boltzmann constant;
T is the temperature;
h is the Planck constant.
The thermal conductivity of nanotubes per unit length is calculated as follows:

$$G = G_{th} N_p \qquad (3)$$

where N_p is the number of phonon channels in a nanotube.
The number of phonon channels N_p is equal to the triple number of atoms in the unit cell $2N$, where N is calculated as follows:

$$N = \frac{2(n^2+m^2+nm)}{d_R} \qquad (4)$$

where d_R is equal to the greatest common divisor of the numbers $(2n + m)$ and $(n + 2m)$.
According to [3, 4] a single-walled carbon nanotube with the chiral indices (5, 5) and the diameter D = 0, 678 nm contains 60 phonon channels and its thermal conductivity is given as follows:

$$G = 60\ G_{th} \qquad (5)$$

A single-walled carbon nanotube with the chiral indices (10, 10) and the diameter D = 1,356 nm contains 120 phonon channels and its thermal conductivity is given as follows:

$$G = 120\ G_{th} \qquad (6)$$

Table I summaries the data for the thermal conductivity of two nanotubes.
It should be noted, that experimental data for the thermal properties of nanotubes has been published in [4]. In particular, the experimental value of thermal conductivity of the nanotube (10, 10) is equal to

$$G = 128\ G_{th} \qquad (7)$$

Through inspecting other experimental data points [4], we can validate that the measured values lie very close to the curves proposed by (3).

3. Thermal properties carbon nanotube framework

We developed a framework for analysis and visualization of nanotubes transport properties [10]. Our framework has been implemented using C# programming language and Microsoft Visual Studio 2015. We use the libraries of Tao Framework, including the OpenGL library [11].
Our framework provides a comparison of thermal properties of two different nanotubes and supports nanotubes variation analysis. More details about the first feature can be found in [12].
In the paper, our focus is on nanotube variation visualization and its analysis.

IOP Conf. Series: Journal of Physics: Conf. Series **829** (2017) 012008 doi:10.1088/1742-6596/829/1/012008

Table 1. The thermal conductivity of nanotubes.

	Indices		Temperature, K		
	n	*m*	233	273	333
The thermal conductivity of the nanotube, *G*, W/K	5	5	1,32 10^{-8}	1,54 10^{-8}	1,89 10^{-8}
The thermal conductivity of the nanotube, *G*, W/K	10	10	2,64 10^{-8}	3,09 10^{-8}	3,78 10^{-8}

3. 1. Nanotubes variation analysis

Nanotubes parameters vary in diameter and chirality indices. In our framework, we provide three different types of diameters variation settings:

1. Min / Max (Figure 1).
2. Deviation, %.
3. Deviation, nm.

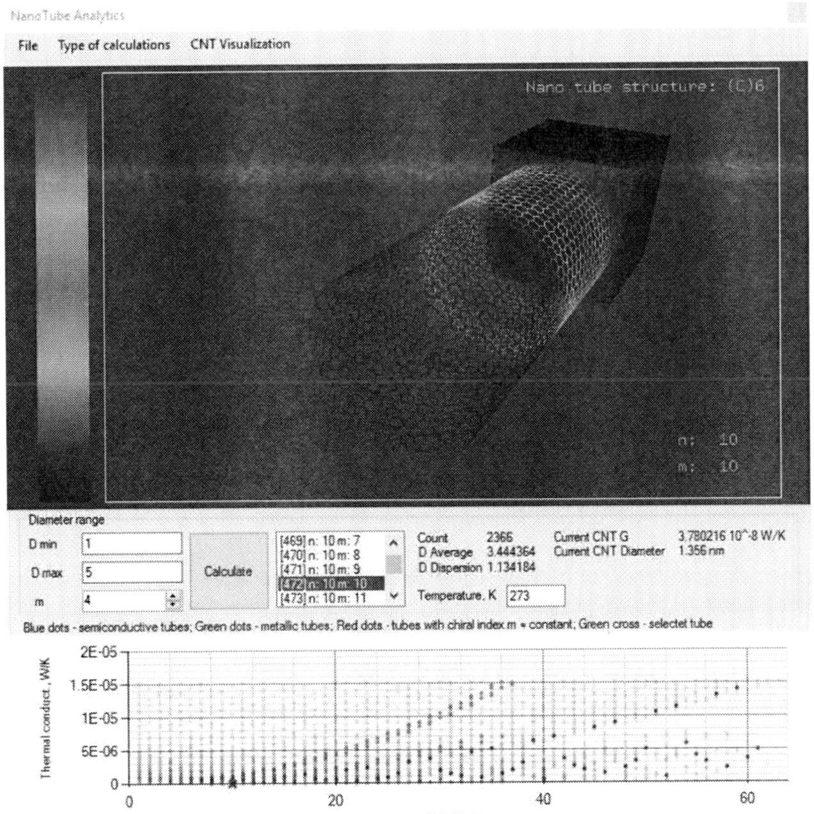

Figure 1*. The variation tube analysis mode. Diameters vary from 1 nm to 5 nm, m = 10. × depicts experimental measures [4] for the nanotube (10, 10).*

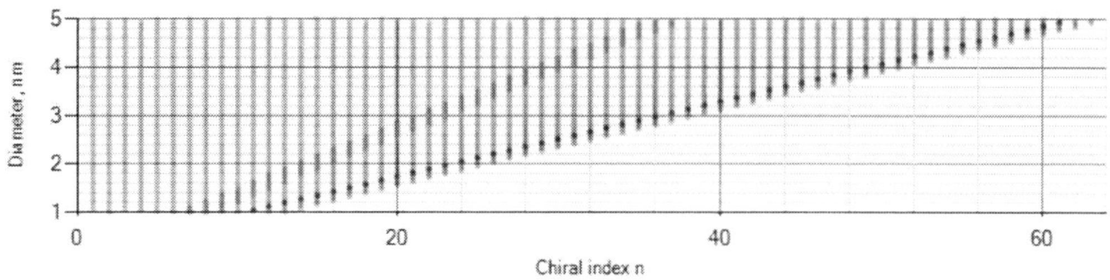

Figure 2. Diameters variation for different chiral indices n.

For all settings, our framework calculates an average nanotube diameter and a nanotube diameter dispersion.

Our framework generates the following output information:

1. Nanotubes diameters variation.
2. The sorted nanotubes diameters.
3. Dispersion of nanotubes diameters.
4. A curve $D(n, m)$.
5. Chiral indices variations.
6. A curve $G(D)$.
7. A curve $G(n, m)$
8. A dispersion of nanotube thermal conductivity.

For a given tube, our framework shows the corresponding thermal conductivity, the diameter and the dot in the curve.

Figure 1 illustrates the variation nanotubes analysis mode of our framework. The curve shows a change of thermal conductivity for different chirality indices. Each blue dot corresponds to semiconductor dots. Green dots show metallic nanotubes. Transparency of each point allows us to estimate the density of the distribution of values in different areas. Figure 1 shows the case $m = 10$.

Figure 2 summaries a diameter variation for variation of the chiral index n, while Figure 3 shows all possible chirality indices n and m for the diameter variations from 1 nm to 5 nm.

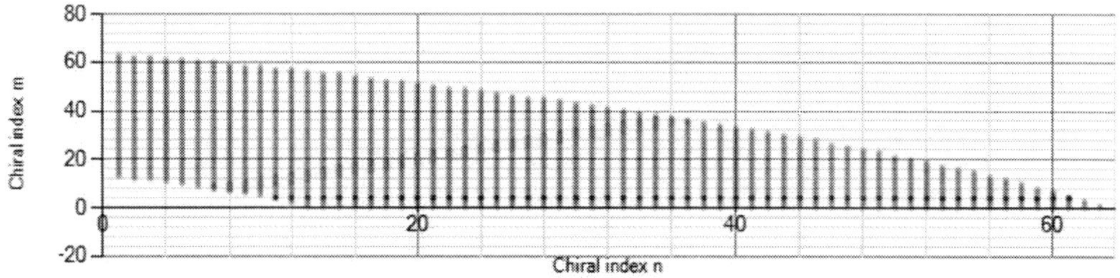

Figure 3. Chirality indices variations for the diameter variations from the 1 nm to 5 nm.

Figure 4 shows thermal conductivity variation for diameter variations. It is obvious that tubes divided into clusters. Tubes with equal m are observed in different clusters. It is remarkable that metallic and semiconductor nanotubes are observed in each cluster.

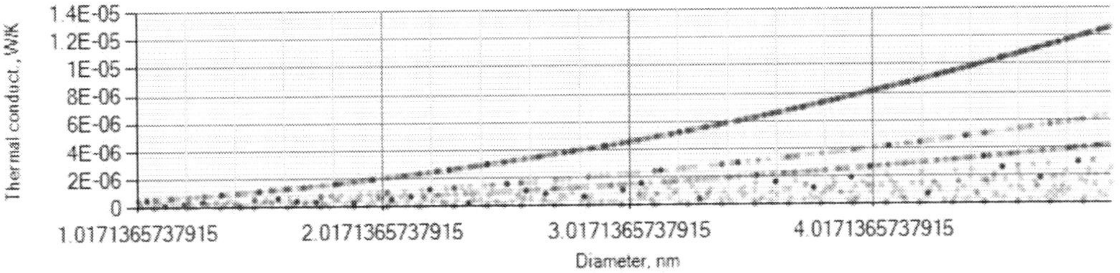

Figure 4. *Thermal conductivity variations*

Figure 5 represents a dispersion of thermal conductivity for nanotube diameter variation from 4 nm to 5 nm. It is remarkable, that some values of thermal conductivity are missing.

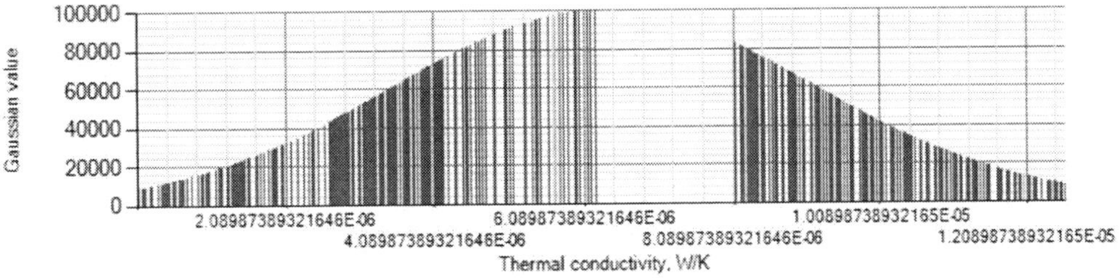

Figure 5. *Thermal conductivity dispersion.*

4. Conclusion

In the paper, we have presented our approach to visual analytics support for investigation of carbon nanotube variations. We have illustrated our approach using the single walled carbon nanotubes thermal conductivity as our test case. Our tool visualizes variation of nanotubes parameters and performance. It is obvious that visual analytics is useful for investigation of thermal properties of carbon nanotubes.

It should be mentioned, that our approach uses only the visual information channel. However, the approach simplifies a nanotube design flow.

Acknowledgments

This work was partially supported by grant RFBR 15-29-01115 ofi-m.

References

[1] Brazhe R, Nefedov V 2014 Thermal conductivity of planar and nanotubular supracrystalline structures at temperatures below the Debye temperature. *Physics of Solid State*, vol **56**, no. 3. pp. 626-630.

[2] Brazhe R, Nefedov V 2012 Thermal conductivity of carbon supracrystalline nanotubes. *Physics of Solid State*, **54**, no. 7. S. 1435-1438.

[3] Eletskii A 1997 Carbon nanotubes. *Advances in Physical Sciences*, 1997, **9**.

[4] Eletskii A 2009 Transport properties of carbon nanotubes. *Advances in Physical Sciences*, **179**, **3**, p. 225-242.

[5] Rieth M, Schommers W 2006 Handbook of Theoretical and Computational Nanotechnology. *American Scientific Publishers*.

[6] Shakhnov V, Zinchenko L, Rezchikova E 2014 Simulation and visualization in cognitive nanoinformatics. *International Journal of Mathematics and Computers in Simulation*, **1** pp. 141-147.

[7] Hills G, et al. 2015 Rapid Co-Optimization of Processing and Circuit Design to Overcome Carbon Nanotube Variations. *IEEE Transactions on Computer-Aided Design of Integrated Circuits and Systems*, **7**, pp. 1082-1095.

[8] Paul B, et al. 2007 Impact of a Process Variation on Nanowire and Nanotube Device Performance. *IEEE Transactions on Electron Devices*, **9**, pp. 2369-2376.

[9] Marconnet A, Panzer M, and Goodson K 2013 Thermal conduction phenomena in carbon nanotubes and related nanostructured materials. *Rev. Mod. Phys.* **85**, no. 8, pp. 1296 - 1327.

[10] Shakhnov V, Zinchenko L, Rezchikova E, Verstov V, Makarchuk V, Sorokin B, Kazakov V 2016 The use of cognitive information and communication technologies in the study of the transport properties of carbon nanotubes. *Design and technology of electronic means*, **1** pp. 8-13.

[11] Kazakov V, Verstov V, Zinchenko L, Averyanikhin A 2016 Problems of WPF applications for Visualization of Transport Properties of Carbon Nanotubes. *Technology engineering and information systems*. **2**. pp. 34-41.

[12] Kazakov V, Verstov V, Zinchenko L, Makarchuk V 2016 Visual Analytics Support for Carbon Nanotube Design Automation. Biologically Inspired Cognitive Architectures (BICA) for Young Scientists. Volume **449**. pp 71-78.

Ca $(OH)_2$Nanoparticles Based on Acrylic Copolymers for the consolidation and protection of Ancient Egypt Calcareous Stone Monuments

Mohammad A. Al-Dosari[1], Sawsan S. Darwish[2], Mahmoud A. Adam[2], Nagib A. Elmarzugi[3], Nadia Al-Mouallimi[4], Sayed M. Ahmed[1,5]

[1]National Nanotechnology Research Center, King Abdulalziz City for Science and Technology (KACST), P.O.Box:6086, Riyadh, Saudi Arabia
[2]Department of Conservation, Faculty of Archaeology, Cairo University, Giza,P.O.Box:12613, Egypt.
[3]Faculty of Pharmacy, Tripoli University and BioNano Integration Research Group, Biotechnology Research Center, LARST, Tripoli, Libya, P.O.Box:13100.
[4]Saudi Nano Information Center, Jeddah P.O.Box:20718, Saudi Arabia.
[1,5]The grand Egyptian Museum, Ministry of Antiquities.Cairo,Egypt.
E-mail:aldosari@kacst.edu.sa, sayedmansour32@yahoo.com,

Abstract. The deterioration of calcareous stones materials used in artistic/architectural field is one of the most serious problems facing conservation today. The aim of this study was to evaluate the effectiveness of nanosized particles of calcium hydroxide (slaked lime) as a consolidation and protection material dispersed in acrylic copolymer, poly ethylmethacrylate/methylacrylate (70:30) (Poly (EMA/MA), for calcareous stone monuments and painted surfaces affected by different kinds of decay. The synthesis process of Ca $(OH)_2$ nanoparticles/polymer nanocomposites have been prepared by in situ emulsion polymerization system. The prepared nanocomposite containing 5% of Ca $(OH)_2$ nanoparticles showed obvious transparency features and represent nanocomposites coating technology with hydrophobic, consolidating and well protection properties.

Keywords: Calcium hydroxide nanoparticles, consolidation, nanocomposites, Compatibility, Ancient Egypt Calcareous Stone Monuments, Conservation.

IOP Conf. Series: Journal of Physics: Conf. Series **829** (2017) 012009 doi:10.1088/1742-6596/829/1/012009

1. Introduction

Limestones from local quarries have been used for the construction of monuments in Egypt from Pharaonic times until today [1]; limestone monuments are suffering from various environmental conditions. The weathering as a natural process is actually one of the most important factor controlling the durability of stone. The deterioration takes the form of loss and granular disintegration of the surface layers of limestone, differential erosion and development of black crusts and orange patinas [2-5].
The water repellence effect for the surface protection of lime stone monument is commonly obtained by the application of polymeric films. Use of water-repellent polymers increases the building materials durability and improves the substrate cleanliness [6, 7]. Acrylic-based compounds have been widely applied for stone consolidation, but with the time, it is worth to mention that the polymer coating appeared, after almost twenty years, quite satisfactory [8, 9]. Therefore, application of nanotechnology to the conservation of artistic and architectural heritage has aroused great interest amongst scientists, conservators and archaeologists and various synthetic routes have been designed to obtain new nanocomposites for conservation purposes [10].
The aim of this study was to evaluate the effectiveness of inorganic compatible treatments, based on nanosized particles of calcium hydroxide (slaked lime) dispersed in acrylic polymer {(poly ethylmethacrylate/methylacrylate (70:30)} in order to improve its properties and compose suitable nanocomposites to be used in the consolidation and protection of limestones and painted surfaces affected by different kinds of decay.

2. Experimental

2.1. Materials

An acrylic copolymer one of the most largely applied copolymers is poly ethylmethacrylate/methylacrylate (70:30) (Poly EMA/MA), commercial name (Paraloid- B72), usually have been widely used, in the treatment of stone artworks and construction materials of historical monuments for consolidation and conservation of such structures [13]. Nanosized particles of calcium hydroxide $Ca(OH)_2$ (with particle mean diameter <50 nm) were purchased from sigma Aldrich.

Nanocomposite preparation by in situ emulsion polymerization system
The $Ca(OH)_2$/polymer nanocomposite has been prepared by different methods as in situ polymerization which was the first method used to synthesis polymer/nanocomposites based on polyamide 66 [14, 15]. The procedure consisted of, synthesis the acrylic polymer with fixed concentration 3% w/v (solid content 3gm/100ml), then the $Ca(OH)_2$ nanoparticles were added during the synthesis of the polymer with concentration 5%, the $Ca(OH)_2$ nanoparticles concentration depends on polymer solid content (See Table 1).
The nanocomposite preparation processes were carried out as in the order 250-mL round flask, monomer, 50 mL deionized water, 0.2 g KOH, 0.5 g emulsifier sodium dodecyl sulfate (SDS), $Ca(OH)_2$ nanoparticles were added with concentration (5%) and stirred for 30 min at room temperature, then the mixture was heated to 80°C. After that the initiator (SPS/SBS) added to the mixture under continuous stirring for 3 h. After cooling, the product precipitated in methanol 1:3(1% methanol and 3% nanocomposite). The precipitated nanocomposites hybrid filtered and washed with methanol. Then dried under vacuum for 24 h at 60°C.

Table 1: Concentrations of used consolidation materials

Consolidation material	Ca $(OH)_2$ nanoparticles concentration	Ca $(OH)_2$ solid content	Polymer solid content	The obtained nanocomposite
Paraliod B-72	-	-	100 ml/3gm	Zero composite
Ca $(OH)_2$ nanoparticles	5%	0.15gm	100 ml/3gm	Ca $(OH)_2$nanoparticles / polymer nanocomposites (5%)

2.2. Preparation of stone samples and application of protective materials

The limestone blocks (samples) as one of the most type of the calcareous stones commonly used for monuments were collected from the quarry of Mokattam limestone plateau east of Cairo city, one of the most important quarries of limestone in Egypt. The specimens were squared blocks, and cut following the UNI10859:2000 [16], then dried in an oven at 105 °C for 24 h, and before the application of the treatments, the stones were kept at 23±2 °C, 50 ± 5% R.H. for 24 h [17].

The pure acrylic polymers and nanocomposites were applied on the limestone samples by brush (three applications). Treated samples were left for 1 month at room temperature and controlled RH 50% to allow the polymerization process to take place. Some of treated samples were submitted to investigation methods and the others were submitted to the artificial aging and then to the investigation methods to monitor the changes of protective materials after accelerated aging test.

2.3. Thermal-aging cycles

Thermal-aging was selected to simulate the natural conditions; the treated samples were put in a temperature-controlled oven "Herous-Germany" on special frames. The samples were thermally aged separately at temperature of 105oC for 24 hours, then the samples output from the oven and immersion in water for 24 hours; this process was repeated 30 cycles [18].

2.4. SEM

SEM Philips (XL30), was used to evaluate the distribution and behaviour of the protective materials on the treated samples and treated aged samples. Images were acquired in backscattered mode (BSE).

2.5. Colorimetric measurements

Colorimetric tests were carried out using a CM-2600d Kon-ica Minolta spectrophotometer to assess chromatic variations before and after treatment and after thermal aging. Chromatic values are expressed in the CIE L*a*b* space, where L* is the lightness/darkness coordinate, a* the red/green coordinate (+a* indicating red and −a* green) and b* the yellow/blue coordinate (+b* indicating yellow and −b* blue) [19].

2.6. Mechanical properties measurement

The measurement of compression strength was carried out before and after treatment and after thermal aging, using an Amsler compression-testing machine, three weeks after treating samples with acrylic polymer and nanocomposite. The average values of compression strength were recorded [20].

IOP Conf. Series: Journal of Physics: Conf. Series **829** (2017) 012009 doi:10.1088/1742-6596/829/1/012009

2.7. Water absorption
The water absorption measurements were carried out using the gravimetric method [21].The limestone samples were completely immersed in deionized water at room temperature. After 24hours, the samples were taken out, wiped with tissue paper carefully and weighed immediately.
The amount of the absorbed water was calculated using the following equation:

$$\text{Water absorption} = \frac{(W_2 - W_1)}{W_1} \times 100 = \cdots\cdots \%$$

Where (W2) is the mass of the sample after immersion in water for 24 hours, and (W1) is the mass of the sample before immersion.

3. Results and discussion

3.1. Scanning electron microscope (SEM) investigations
The examination by SEM was used to study the ability of consolidation materials to consolidate and protect the limestone samples. The SEM micrographs of the untreated sample, (figure 1a), appeared to be very fragile, suffered from granular disintegration, and presence of some voids was noticed because of dissolving and disappearance of binding materials. The SEM examination of the samples treated with pure Paraloid B-72(figure1b), showed that, the consolidant filled most of the pores and obscured many of particles, however material has failed to fill the fine cracks in some depth areas. On the contrary, the samples treated with Ca $(OH)_2$ /polymer nanocomposites (figure 1C), showed a homogeneous coating of the particles more than the coating with Paraloid B-72 without nanoparticles, no visible cracks are seen in the film of coating, this may be due to the fact that the higher physical-chemical and physicalmechanical compatibility with the original materials as well as the possibility to penetrate deep into damaged zones without limitations due to the particle size and better stability. After artificial thermal aging, small changes were observed in samples treated with both products, the film uniform and homogeneity was affected as a result of aging, which was illustrated further in samples treated with Paraliod B-72(figure 1d), On the other hand, it was found that the addition of nanoparticles to the polymers (figure 1e), improve the stability of microstructure of the product under the effect of the artificial thermal aging compared to those coated with Paraliod – B72 without the nanoparticles.

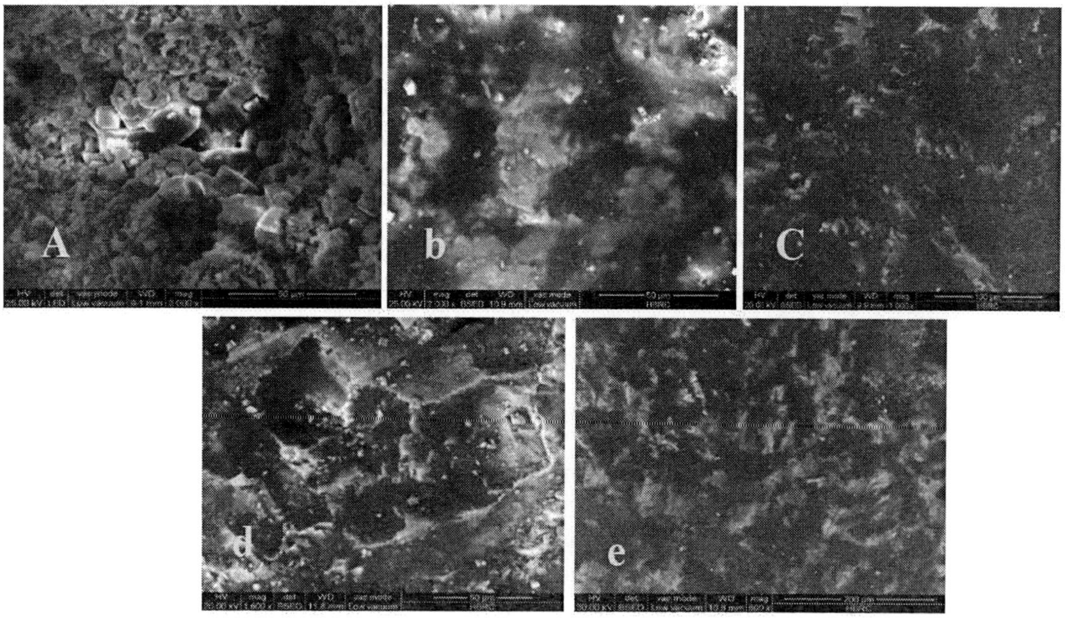

IOP Conf. Series: Journal of Physics: Conf. Series **829** (2017) 012009 doi:10.1088/1742-6596/829/1/012009

Figure.1. Shows the SEM micrographs of (A) untreated sample, (b). Sample treated with Pure Paraliod – B72, (c). Sample treated with Ca (OH) 2 /polymer nanocomposites, (d). Sample treated with Paraliod – B72 after thermal aging and (e). Sample treated with Ca (OH) $_2$ /polymer nanocomposites after thermal aging.

3.2. Mechanical Properties

The mechanical properties of the untreated, treated and treated samples after artifial aging were determined by testing the compressive strength. Table 2 shows the average values of compressive strength for limestone samples. By comparison, it was found that the addition of nanoparticles to the polymer increase its compressive strength values. This may be attributed to the role of nanoparticles in reinforcing the polymer, and also improving its interaction with the stone grains.

Table 2. Average values of compressive strength for untreated, treated and treated aged limestone samples

Consolidation material	Compressive strength Cm^2/ Kg	
	After treatment	After treatment and thermal aging
Untreated samples	210	---
Paraloid B72	230	215
Ca $(OH)_2$/Polymer nanocomposites	245	235

3.3. Color measurement

The chromatic changes ΔE*ab were also carried out by means of Optimatch 3100, in order to calculate and determine the variation of the aesthetical properties induced by the treatments, according to the following equation:

$$\Delta E = \sqrt{\Delta L^{*2} + \Delta a^{*2} + \Delta b^{*2}}$$

Where ΔL*, Δa* and Δb* are the differences in the, L*, a* and b* coordinates (according to CIEL AB color space) of the treated and untreated limestone samples. According to Italian guidelines for the restoration of stone monuments, the ΔE value must be < 5, After treatment and aging, negligible color variations were observed, and all values are in acceptable limit (ΔE value < 5), thus confirming the suitability of the product for restoring purposes. The obtained data are fully listed in Table.3

Table 3. Color measurement in treated and aged samples

Applied consolidation materials	Δ (treated and untreated samples)				Δ (thermal aged and untreated samples)			
	ΔL^*	Δa^*	Δb^*	ΔE	ΔL^*	Δa^*	Δb^*	ΔE
Paraloid B72	- 1.08	- 0.38	0.79	1.39	2.03	0.41	0.66	2.17
Ca $(OH)_2$/Polymer nanocomposites	3.39	0.48	1.85	3.89	3.61	0.54	1.74	4.04

3.4. Water absorption

Since the water is considered to be the major deterioration factor, it is very important the materials of consolidation and protection are able to reduce water penetration into the stone bulk. It is evident, from the physical measurements, that the treated samples are higher in their bulk density. By measuring the water absorption values of the samples treated with pure polymers and nanocomposites, it was found that addition of nanoparticles to the polymers led to reduce their water absorption rates. The efficiency of the nanocomposite in formation of a protective layer appears from the reduction in water absorption and porosity, which can be referred to the penetration of the nanoparticles into voids and pores, also this is attributed to the improving of physiochemical properties of the polymers by nanoparticles, which also led to decreasing the cracking rates during the drying process [22,23]. Table 4 shows the average values of water absorption for the treated and untreated sandstone samples.

Table 4. Average values of water absorption for treated and aged lime stone samples

Consolidation material	Water absorption%	
	After treatment	After treatment and thermal aging
Untreated Samples	4.8 %	---
Paraloid B72	3.9 %	4 .3 %
Ca $(OH)_2$/Polymernanocomposites	1.769	2.02 %

4. Conclusion

In this study, $Ca(OH)_2$/polymer nanocomposites were successfully prepared by in situ emulsion polymerization system, to produce coating with high transparency. Adding of Ca $(OH)_2$ nanoparticles to poly ethylmethacrylate/methylacrylate (70:30) (Poly (EMA/MA) in order to improve its physiochemical and mechanical properties, in order to use it in the consolidation and protection of limestones. Samples treated with (Poly (EMA/MA), and Ca $(OH)_2$/polymer nanocomposites; were tested before and after treatment and under artificial thermal aging; the results showed that the addition of nanoparticles to acrylic polymer improved its ability to consolidate and protect the limestone samples. The polymer containing Ca $(OH)_2$ nanoparticles have the great advantage of high compatibility with the constituting materials of limestone samples, which achieve a deeper penetration of the dispersion, better stability and avoid the formation of white glazing on the treated surface. Moreover, the addition of Ca $(OH)_2$ nanoparticles enhances the physical and mechanical properties of the polymer and treated samples, compared to those treated with polymer without the nanoparticles.

References

[1] `Saleh, S. A., Helmi, F. M., Kamal, M. M. & El-Banna, A. (1992). Study and consolidation of sandstone: Temple of Karnak, Luxor, Egypt, Studies in Conservation. Paper presented at the *The Proceedings of the Seventh International Congress on Deterioration and Conservation of Stone*, Lisbon, Portugal.

[2] Hawass, Z. (1995) The Egyptian Monuments: Problems and Solutions, *International Journal of Cultural Property.* **4**, 105-118.

[3] Croci, G. (1996) Damage and Restoration of Monuments in Cairo in *Protection and Conservation of the European Cultural Heritage Research Report n° 4* (Zezza, F., ed) pp. 425-431, University C.U.M. School Monument Conservation, Bari, Italy.

[4] Fitzner, B., Heinrichs, K. & Bouchardiere, D. L. (2003) Weathering Damage on Pharaonic Sandstone Monuments in Luxor-Egypt, *Building and Environment.* **38**, 1089-1103.

[5] Warscheida, T. & Braamsb, J. (2000) Biodeterioration of stone: A review, *International Biodeterioration & Biodegradation.* **46**, 343-368.

[6] Poli, T., Toniolo, L. & Chiantore, O. (2004) The protection of different Italian marbles with two partially flourinated acrylic copolymers, *Applied Physics A.* **79**, 347-351.

[7] Rizzarelli, P., La Rosa, C. & Torrisi, A. (2001) Testing a fluorinated compound as a protective material for calcarenite, *Journal of Cultural Heritage.* **2**, 55-62.

[8] Garrido, J. M. (1967) The portal of the monastery of santa maria de ripoll, *Monumentum.* **1**, 79-98.

[9] Ditaranto, N., Loperfido, S., van der Werf, I., Mangone, A., Cioffi, N. & Sabbatini, L. (2011) Synthesis and analytical characterisation of copper-based nanocoatings for bioactive stone artworks treatment, *Analytical and Bioanalytical Chemistry.* **399**, 473-481.

[10] Baglioni, P., Giorgi, R. & Chelazzi, D. (2012) Nano-Materials for the Conservation and Preservation of Movable and Immovable Artworks, *International Journal of Heritage in the Digital Era.* **1**, 313-318.

[11] Giorgi, R., Dei, L. & Baglioni, P. (2000) A New Method for Consolidating Wall Paintings Based on Dispersions of Lime in Alcohol *Studies in Conservation.* **45**, 154-161.

[12] Ambrosi, M., Dei, L., Giorgi, R., Neto, C. & Baglioni, P. (2001) Colloidal Particles of Ca(OH)2: Properties and Applications to Restoration of Frescoes, *Langmuir.* **17**, 4251-4255.

[13] Vaz, M. F., Pires, J. & Carvalho, A. P. (2008) Effect of the impregnation treatment with Paraloid B-72 on the properties of old Portuguese ceramic tiles, *Journal of Cultural Heritage.* **9**, 269-276.

[14] Effenberger, F., Schweizer, M. & Mohamed, W. S. (2009) Synthesis and characterization of some polyacrylate/montmorillonite nanocomposites by in situ emulsion polymerization using redox initiation system, *Journal of Applied Polymer Science.* **112**, 1572–1578.

[15] Zhuang, W., Liu, J., Zhang, J. H., Hu, B. X. & Shen, J. (2008) Preparation, characterization, and properties of TiO2/PLA nanocomposites by in situ polymerization, *Polymer Composites.* **30**, 1074–1080.

[16] UNI 10859:2000, Cultural Heritage - Natural And Artificial Stones - Determination of Water Absorption By Capillarity in *Mineral materials and products*, Unifica zione Italian no(UNI), Italy.

[17] Licciulli, A., Calia, A., Lettieri, M., Diso, D., Masieri, M., Franza, S., Amadelli, R. & Casarano, G. (2011) Photocatalytic TiO2 coatings on limestone, *Journal of Sol-Gel Science and Technology.* **60**, 437-444.

[18] Henriques, D. F., Nunes, L. & Brito, J. d. (2013) Mechanical evaluation of timber conservation processes by bending tests, *Advanced Materials Research.* **778**, 612-619.

[19] Quagliarini, E., Bondioli, F., Goffredo, G. B., Licciulli, A. & Munafò, P. (2013) Self-cleaning materials on Architectural Heritage: Compatibility of photo-induced hydrophilicity of TiO2 coatings on stone surfaces, *Journal of Cultural Heritage.* **14**, 1-7.

[20] Standard, A. Method of Test for Compressive Strength of Natural Building Stone (98).

[21] Luvidi, L., Mecchi, A. M., Ferretti, M. & Sidoti, G. (2012) Evaluation of Tio2 Based Treatments on Stone Surfaces Exposed to A Polluted Urban Environment in *12th International Congress on the Deterioration and Conservation of Stone* (Wheeler, G., ed), Columbia University, New York.

[22] Khallaf, M. K., El-Midany, A. A. & El-Mofty, S. E. (2011) Influence of acrylic coatings on the interfacial, physical, and mechanical properties of stone-based monuments, *Progress in Organic Coatings.* **72**, 592-598.

[23] de Ferri, L., Lottici, P. P., Lorenzi, A., Montenero, A. & Salvioli-Mariani, E. (2011) Study of silica nanoparticles – polysiloxane hydrophobic treatments for stone-based monument protection, *Journal of Cultural Heritage.* **12**, 356-363.

Enhancement of electrochemical properties of micro/nano electrodes based on TiO_2 nanotube arrays

[1]D Khudhair, [2]J Gaburro, [3]S Shafei, [4]A Barlow, [1]S Nahavandi and [1]A Bhatti

[1]Institute for Intelligent Systems Research and Innovation, Deakin University, Vic 3216, Australia
[2]Australian Animal Health Laboratory - CSIRO, Vic 3219, Australia
[3]Institute for Frontier Materials, Deakin University, Vic 3216, Australia
[4]Center for material and surface science, Department of chemistry and physics, La Trobe University, Vic 3086, Australia

E-mail: dkhudhai@deakin.edu.au

Abstract. Titanium oxide nanotube (TiO_2 nanotube) arrays were produced by anodizing titanium foils in two different electrolytes. The first electrolyte consisted of ethylene glycol containing 0.5 wt% NH_4F and 4 vol% of distilled water to produce pure TiO_2 nanotube arrays and the second consisted of HF aqueous solution (0.5 wt%) containing 0.5% polyvinylalcohol to produce carbon doped TiO_2 nanotube arrays. The fabricated TiO_2 nanotube arrays were subsequently annealed in the atmosphere of nitrogen. The morphology and crystal structure of fabricated arrays were characterized by means of scanning electron microscopy and X-ray diffraction. The electrical conductivity and capacitance of TiO_2 nanotube arrays were investigated by electrochemical impedance spectroscopy (EIS)and cyclic voltammetry (CV). Water contact angle and biocompatibility of fabricated nanotube arrays were investigated. The results showed that carbon doped TiO_2 nanotube arrays annealed in the atmosphere of nitrogen have higher conductivity and capacitance than those of pure arrays annealed in the same atmosphere. Doping with carbon enhances the biocompatibility and wettability of TiO_2 nanotube arrays. It has also noted that electrical conductivity and capacitance of TiO_2 nanotube arrays were directly proportional to the tube wall thickness.

1. Introduction

Recently, considerable attention has been paid to TiO_2 nanotube arrays as they can provide a direct way for charge transport and provide a large surface area. TiO_2 nanotube arrays were fabricated using various approaches, for example, template, anodization and hydrothermal synthesis [1]. Anodization is the most common process due to its low cost, it is also an environmentally friendly and highly controllable process by which highly ordered arrays can be fabricated with different tube morphologies. These highly ordered arrays have a great prospect to be employed as electrochemical electrodes [2]. However, the low conductivity of pure form and amorphous nature of TiO_2 nanotube arrays resulting from anodization is a significant obstacle that limits them for such application [3]. To overcome this problem, chemical and physical approaches such as annealing in different gases, anodization and chemical deposition were employed to introduce transition metals such as Pt, Cr and Ag or nonmetal elements including C and N on TiO_2 lattice [4]. TiO_2 can exist in three different crystalline structure

Table 1. diameter, wall thickness and length of pure and carbon doped TiO_2nanotubes annealed in atmosphere of nitrogen.

TiO_2 nanotube array	Tube diameter (nm)	Tube wall thickness (nm)	Tube length (μm)
TNT-25V	37.68	13.52	1.38
TNT-C-15V	38.23	13.61	0.791
TNT-35V	57	18.45	2
TNT-C-20V	56.89	17.96	1.16
TNT-45V	68	23.2	3.7
TNT-C-25V	68.75	2.43	2.2

forms; rutile, anatase and brookite. In general, transformation from amorphous to crystalline form occurs when TiO_2 is heated above 350 °C when anatase is formed, while rutile is the most stable phase at temperatures greater than 800 °C [5, 6]. It has been shown that brookite is a rare phase that could exist as a minor phase depending on fabrication conditions [7]. Among the three crystalline structures, anatase is shown to be the most conductive phase [8]. In this work, anodic oxidation process was employed to fabricate pure and carbon doped TiO_2 nanotube arrays with convergent tube diameter and wall thickness. The results showed that TiO_2 nanotube arrays doped with carbon and annealed in the atmosphere of nitrogen have higher electrical conductivity and capacitance than those of pure TiO_2 nanotube arrays annealed in the same atmosphere. Carbon doped nanotube arrays showed higher biocompatibility and wettability than those of pure nanotube arrays.

2. Experimental section

Six rectangular samples of titanium foils (purity of 99.7%) with dimension of 1 cm x 1.5 cm were cleaned by ultrasonic treatment in solution of ethanol and de-ionized water for 10 minutes. An electrochemical cell of platinium cathode was used to anodize the titanium samples which were placed as anode electrodes and separated by 2 cm from the cathode. Three samples were anodized in electrolyte consisting of ethylene glycol (anhydrous, 99.8% sigma Aldrich) containing 0.5 wt% NH_4F and 4 vol% of distilled water (electrolyte A) at voltages of 25 V, 35 V and 45 V for one hour. An electrolyte of HF aqueous solution (0.5 Wt%) containing 0.5 wt% polyvinylalcohol with molecular weight of 89000-98000 (electrolyte B) was used to anodize the other samples at applied voltages of 15 V, 20 V and 25 V for 25 minutes. The experiments were conducted at 20°C. All anodized samples were annealed in a nitrogen atmosphere at 500 °C for 3 hours. X- ray diffraction (XRD, Panalytical X-Pert Pro MRD XL,CuKα radiation,λ=1.5418 Å) was employed to investigate the phase and crystal structure of the samples. The XRD measurements were conducted within 2θ range from 20° to 60°. Match software was used to analyse the recorded pattern. To confirm the existence of carbon atoms in the oxide film, X-ray photoelectron spectroscopy test (XPS) was done on C-doped nanotube arrays. Kratos AXIS Nova instrument, which utilises a monochromatic Al K X-ray source (1486.69 eV) was used for XPS test. Scanning electron microscopy (SEM, Carl Zeiss SUPRA 55VP instrument) was used to observe the morphology of TiO_2 nanotubes. The electrochemical measurements were carried out in a three electrode cell. A platinum standard electrode was the counter electrode, Ag/AgCl was the reference electrode and TiO_2 nanotubes on the sample was the working electrode. For electrochemical tests, the electrochemical cell was filled with potassium buffered saline solution. The EIS was measured at a frequency of 1 KH. A range of potential from -0.1 V to 0.6 V versus saturated Ag/AgCl was applied in the cyclic voltammetry test. The area of the sample which was exposed to the solution of the electrochemical cell was (4.9 mm^2) in all electrochemical experiments. To evaluate the wettability, water contact angle was

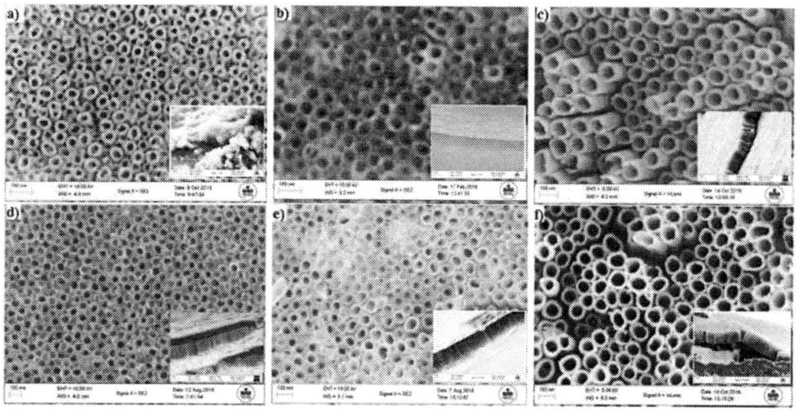

Figure 1. SEM images of TiO_2 nanotube arrays of (a) TNT-C-15 V, (b) TNT-C-20 V, (c) TNT-C-25 V, (d) TNT-25 V, (e) TNT-35 V and (f) TNT-45 V.

measured for all anodized samples. The water drop volume was 3 ml. All measurements were conducted at room temperature. To conduct indirect cytotoxicity test, VERO cells (ATCC CCL81 African green Monkey Kidney) seeded into a media previously incubated in contact with the samples. Samples were autoclaved for 45 minutes at 121 °C, for sterilization. The utoclaved samples were placed in a 24-wells plate, dried and coated with PEI (polyethylenimine). The PEI was removed and 30 μl of laminin was spread over the surface of each pre-coated sample. Then, The plate was incubated in the atmosphere of 2% CO2 at a temperature of 37 C for 30 minutes. Laminin was removed and 1 ml of the media containing VERO cells was added to each sample. Cells on nanotube samples were let to rest in the incubator for 2, 4 and 7 days. For each appropriate time after cells seeding, samples were removed and placed into a new plate for cells counting. First the samples were rinsed with Phosphate-Buffered Saline solution (PBS) and then followed by a Trypsin treatment to separate cells from the sample surfaces. Cells were manually counted and estimated using Trypan blue staining method on a Hemocymeter. Scanning electron microscopy was used to observe the morphology of the cells cultured on the TiO_2nanotube arrays. After the cell culture, cells were fixed in 4% paraformaldehyde for 10 min. Followed by 3 times washing using PBS, cells were then dehydrated by increasing concentrations of ethanol in water (25%, 50%, 70%, 90%, 100% and again 100% for 5 minutes each). The dried cells on the electrodes were placed in a vacuum chamber overnight and images from each sample were captured using Zeiss Supra 55VP to evaluate the cell coverage on each sample. Note that characterization of nanotube morphology, electrochemical test and biological test were performed on fabricated nanotube arrays after annealing in atmosphere of nitrogen.

3. Results and discussion

pure TiO_2 nanotube arrays with three various morphologies were fabricated by anodizing three samples at 25 V, 35 V and 45 V in electrolyte A (group 1). To obtain C-doped TiO_2 nanotube arrays with morphologies close to those of pure nanotubes, three other samples were anodized in electrolyte B at 15 V, 20 V and 25 V (group 2). For convenience, the nanotubes arrays which were fabricated in electrolyte A at voltages of 25 V, 35 V and 45 V were named as TNT-25 V, TNT-35 V and TNT-45 V respectively, while nanotubes arrays which were produced in electrolyte B at 15 V, 20 V and 25 V were named as TNT-C-15 V, TNT-C-20 V and TNT-C-25 V respectively. It was observed that the pairs (TNT-25V, TNT-C-15V), (TNT-35V, TNT-C-20V) and (TNT-45V, TNT-C-23V) were convergent in terms of tube diameter and tube wall thickness, while nanotube

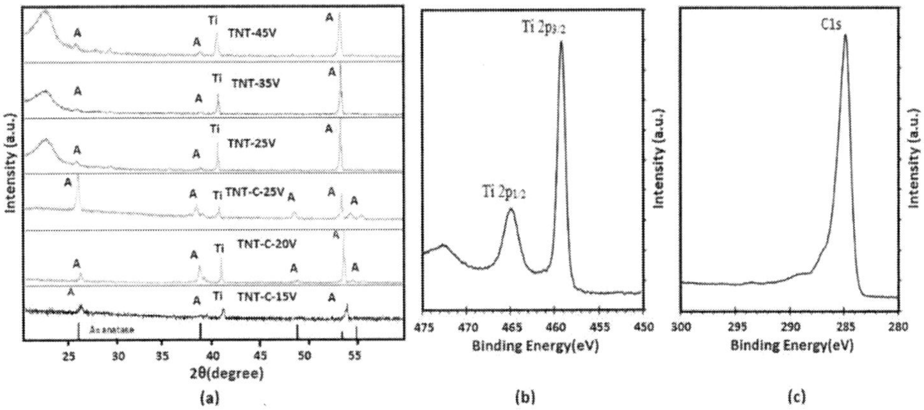

Figure 2. (a) X-ray diffraction patterns of fabricated TiO_2 nanotube arrays (b) XPS spectrum (Ti2p) for C-doped nanotube arrays (c) XPS spectrum (C1s) for C-doped nanotube arrays .

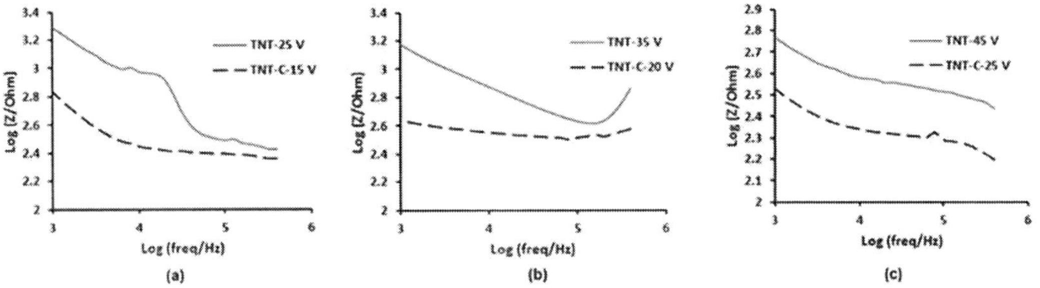

Figure 3. Bode plots for (a) TNT-C-15V and TNT-25V (b) TNT-C-20V and TNT-35V (c) TNT-C-25V and TNT-45V.

arrays result in anodization in electrolyte A had a bigger tube length than those anodized in electrolyte B. Figure 1 is the SEM images of the fabricated TiO_2 nanotube arrays. The values of tube diameter, tube wall thickness and tube length are listed in table 1. Figure 2a shows the X-ray diffraction patterns for all annealed samples. It can be seen that TiO_2 nanotube films have main peaks that match with the anatase phase of titanium dioxide at 2θ values of (25.31), (37.57), (47.91), (53.75) and (55.08). In the patterns of undoped samples, peaks appeared in 2θ value of 20 as an instrument noise appeared at the beginning of the experiment. As TiO_2 nanotube arrays result from anodization process have amorphous structures [9], X-ray diffraction patterns give a good indication of successful transformation from amorphous to crystal structure after annealing. To investigate the composition of C-doped TiO_2 nanotube arrays, XPs analysis was performed. Figure 2b shows Ti 2p spectra with two peaks obtained corresponding to $Ti2p_{3/2}$ and $Ti2p_{1/2}$ at 459.2 eV and 465 eV respectively indicating that this signals corresponded to TiO_2 [10]. The peak C1s appeared at 284.6 eV which could be attributed to C-Ti [11]. According to Valentin at el the presence of carbon atoms in the lattice of anatase TiO_2 culd be attributed to the replacement of some of oxygen and titanium atoms with carbon atoms [12]. The electrochemical impedance plots of the three pairs of samples are illustrated in Figure 3. The EIS values of the TiO_2 nanotube arrays of the pairs of samples changed in the following order: 1905.5 Ω (TNT-25V) > 680 Ω (TNT-C-15V) , 1479 Ω (TNT-35V)> 436.5 Ω (TNT-C-20V) and 583.4 Ω (TNT-45V) > 338 Ω (TNT-C-25V). It can bee seen that EIS value

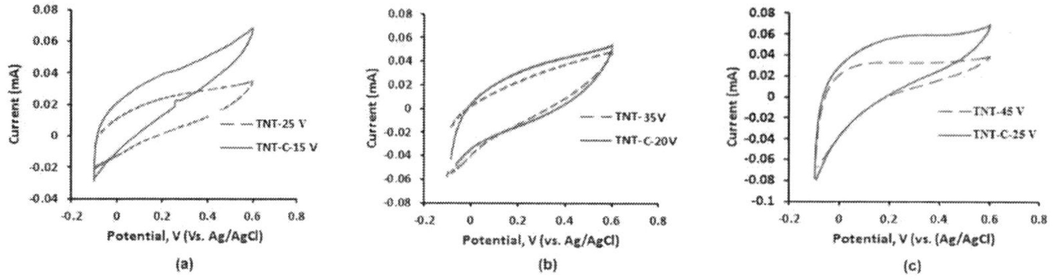

Figure 4. Cyclic voltammetry curves of (a) TNT-C-15V and TNT-25V (b) TNT-C-20V and TNT-35V (c) TNT-C-25V and TNT-45V.

decreases with the increase of tube wall thickness for the samples in both groups. This inverse relationship between tube wall thickness and electrical conductivity could be due to the path of the charge transformation. Thicker tube wall provides greater support for the the space charge layer created through charge movement [13]. In each pair of the samples, TiO_2 nanotube arrays were convergent in tube diameter and tube wall thickness but nanotubes arrays in group 1 have longer tubes. The electrochemical impedance is inversely proportional to the tube diameter and tube wall thickness and directly proportional to the tube length [14]. As nanotubes in group 2 have thicker walls and shorter length, that nanotubes in group 2 have lower EIS than those in group 1 in each pair. The existence of carbon atoms in the lattice of TiO_2 provides a path with lower resistance for charge movement and consequently enhance the electrical conductivity. The Cyclic voltammetry (CV) curves of the six samples of TiO_2 nanotube arrays were shown in Figure 4. The capacitance of annealed nanotubes electrodes were calculated from the recorded (CV) curves using the equation: $C = S/2mk(U1 - U2)$ where C is the capacitance, S is the area between charge and discharge curves, m is the mass of active material, k is the potential scanning rate, $U1$ is the initial potential and $U2$ is the set potential. The obtained values of capacitance were found to be : 56.2 F/mg (TNT-C-15 V) $>$ 37.1 F/mg (TNT-25 V), 109 F/mg (TNT-C-20) $>$ 56.4 F/mg (TNT-35 V) and 130 F/mg (TNT-C-25 V) $>$ 77.1 F/mg (TNT-45V). From the capacitance values, it is clear that in each pair, nanotubes on the samples of group 2 have higher capacitances than those on the samples of group 1. This difference in capacitance values between the samples of the tow groups may be attributed to the difference in electrical conductivity. Samples of group 2 have higher electrical conductivity. This higher conductivity leads to higher enhancement in charge transformation through charge- discharge process and consequently higher capacitance. Water contact angle measurements show that all samples show hydrophilic behaviour. Figure 5-a shows the images of the water drops on the fabricated nanotube arrays. It was observed that contact angle increases with increase of tube diameter. This could be due to a decrease in roughness and effective surface area with the increase in diameter [15]. Figure 5-b illustrate that carbon doped nanotube arrays show more hydrophilicity than pure nanotube arrays having convergent tube diameter and wall thickness. To investigate the biocompatibility, VERO cells were seeded on the fabricated nanotube arrays and cell proliferation test on the fabricated nanotube arrays was performed. As it is shown in figure 6-a, cell density on all nanotube arrays increases with time indicating that the six samples have a good biocompatibility. Figure 6-b is the SEM images of the cultured cells on the sample surfaces after 4 days of cell seeding. SEM images show that in both of pure and C-doped nanotube arrays, cell density decrease with an increase in applied voltage. This could be explained by the increase of tube diameter as a result of the increase in the applied voltage. This observation is in line with the observation in some earlier study [16]. Furthermore, Figure 6(a and b) shows that in each pair of nanotube arrays, C-doped nanotube arrays have higher cell density than those

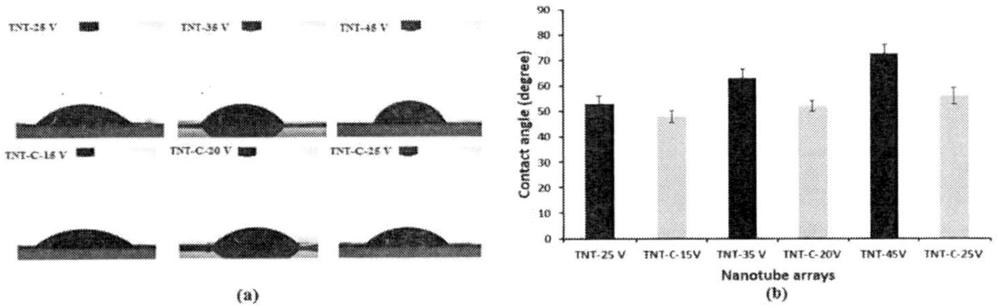

Figure 5. (a) Images of water drop on the surfaces of the fabricated nanotube arrays, (b) contact angle of water drope on the fabricated nanotube arrays. Error bars show standard deviation for n = 5 samples

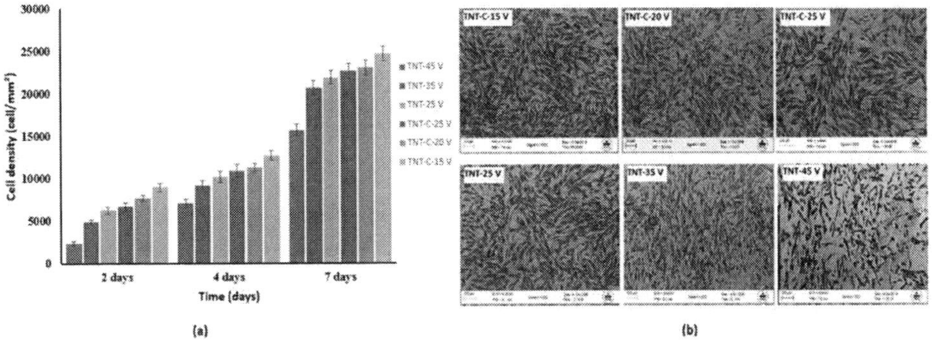

Figure 6. (a) Cell proliferation in terms of cell density measured after 2, 4 and 7 days. Error bars show standard deviation for n = 3 samples, (b) SEM images of cultured cells on the six samples

of pure arrays indicating that carbon doping increases the biocompatibility of TiO_2 nanotube arrays.

4. Conclusions

In this study, we investigate the electrochemical properties, wettability and biocompatibility of TiO_2 nanotube arrays produced by anodic oxidation in two different electrolytes and annealed in atmosphere of nitrogen. The first arrays consist of pure TiO_2 nanotubes and the second include TiO_2 nanotubes doped with carbon. The results of electrochemical measurements showed that the electrical conductivity and capacitance of the nanotubes arrays are directly related to the tube wall thickness. carbon doped TiO_2 nanotubes arrays have lower impedance and higher capacity compared Pure TiO_2 nanotube arrays. Both groups of TiO_2 nanotubes arrays have good biocompatibility and wettability. Water contact angle and biological test revealed that doped with carbon enhances the biocompatibility and wettability of TiO_2 nanotubes arrays.

References

[1] Amani Hamedani H, Lee S, Al-Sammarraie A, Hesabi Z R, Bhatti A, Alamgir F, Garmestani H and Khaleel M 2013 *ACS applied materials and interfaces* **5** 18
[2] Uchida M, Kim H, Kokubo T and Fujibayashi S 2003 *J. Biomedical Materials Research Part A* **64** 1
[3] Khudhair D, Bhatti A, Li Y, Hamedani H, Garmestani H, Hodgson P and Nahavandi, S 2016 *Materials Science and Engineering: C* **59** P 1125-1142
[4] Hamedani H, Allam N, Garmestani H and El-Sayed M 2011 *The Journal of Physical Chemistry C* **115** 27
[5] Richards B 2002 *Center for Photovoltaic Engineering* P 30-37

[6] Bahar A 2011 *University of Baghdad* P 1-5
[7] Popat K, Leoni L, Grimes C and Desai T 2007 *Biomaterials* **28** P 3188–3197
[8] Uchida M, Kim H, Kokubo T, Fujibayashi S and Nakamura T 2003 *J. of Biomedical Materials Research Part A* **64** P 164–170
[9] Morris A and Meyer G 2009 The Journal of Physical Chemistry C **112** P 18224–18231
[10] Choi Y, Umebayashi T, and Yoshikawa M 2004 *Journal of Materials Science* **39**5
[11] Anitha V, Lee J, Banerjee A, Joo S and Min B 2015 *Nanotechnology* **26**6
[12] Cristiana Di Valentin, Gianfranco Pacchioni and Annabella Sellon X 2005 *Chem. Mater* **17** P 6656-6665
[13] Sun L, Zhang S, Sun X and He X 2010 *J. of nanoscience and nanotechnology* **10** P 4551-4561
[14] Xiao P, Zhang, Yunhuai G, Betzaida B and Sepehri S 2009 *J. of nanoscience and nanotechnology* **9** 4
[15] Hanz I, Djenizian T and Knauth P 2011 *The Journal of Physical Chemistry C* **115**3
[16] Park J, Bauer S, von der Mark K and Schmuki P 2007 *Nano letters* **7** 6

IOP Conf. Series: Journal of Physics: Conf. Series **829** (2017) 012011 doi:10.1088/1742-6596/829/1/012011

Effect of Undensified Silica Fume on the Dispersion of Carbon Nanotubes within a Cementitious Composite

S. Alrekabi[1], A. Cundy[2], Raymond L.D. Whitby[3], A. Lampropoulos[1], I. Savina[4],

[1] School of Environment and Technology, University of Brighton, UK
[1]Dept. of Civil Engineering, University of Babylon, Babylon, Iraq
[2] School Ocean and Earth Science, University of Southampton, UK
[3]School of Engineering, Nazarbayev University, Astana, Kazakhstan
[4]School of Pharmacy and Biomolecular Sciences, University of Brighton, UK

Abstract. The synergistic effect of multi-walled carbon nanotubes (MWCNTs) and Undensified Silica Fume (USF) on the microstructure of cementitious composites has been studied. In the current work, USF was used to enhance the dispersion of nanotubes throughout the composite and prevent the re-agglomeration of nanotubes by providing a physical barrier of particles of small size. Ultrasonication was employed to disperse MWCNTs in water in the presence of polycarboxylate-based superplasticizer (PCE) as a dispersion agent. The results indicate that incorporation of USF considerably improves the dispersion of nanotubes in the composites, with subsequent enhancement of composite packing density. This enhancement can be attributed to the synergistic effect of MWCNTs and USF in reducing the volume of pores through the cementitious composites.

Keywords: Cementitious composites, multiwall Carbon nanotubes (MWCNTs), Undensified Microsilica (USF).

1. Introduction

Since the study published Iijima in 1991 [1], carbon nanotubes (CNTs) are emerging as a promising candidate for many applications, however the agglomeration tendency of CNT material presents a considerable barrier to realising the potential offered by these nanostructures [2]. The outstanding properties of carbon nanotubes (CNTs) make them promising candidates as a reinforcing element in cementitious composites [3, 4]. Based on their mechanical and physical properties, CNTs are one of the strongest and stiffest materials known to date. Compared to mild steel, the most widely used reinforcing material for concrete [5], CNTs are about 100 times stronger, 6 times lighter, with a Young's modulus about 5 times greater, and elongation at break up to 60 times greater [5-7].

Similar to other application, dispersion of CNTs for cementitious application is however a highly challenging task, significantly affecting the mechanical properties when added to a composite. The poor wettability of nanotubes and the van der Waals interaction forces between nanotubes create bundles and agglomerates, which hinders their dispersion [8, 9]. Poor dispersion of CNTs in cementitious composites can lead to reduction of mechanical performance, as sufficient dispersion is a prerequisite for their reinforcing and crack-arresting properties [10-12]. In recent works, various approaches including mechanical and physical methods have been proposed to achieve uniform dispersion. The two main dispersion techniques are (i) CNTs added to the cement as a powder and mixed using rotary mixer [13, 14], (ii) ultrasonication technique via covalent/non-covalent bond in combination with a dispersion agent to disperse the nanotubes in aqueous liquid [11, 15]. Moreover, efforts to achieve a proper dispersion through the nano composites were extended to use the pozzolanic admixtures [16-18]. Microsilica has been extensively used as partial substitution of portland cement in cementitious composites, to improve the durability and mechanical properties of the composites[19]. Utilization of microsilica in nano cementitious composites have also been proposed as a means to improve the dispersion of nanotubes in composites. Their great effect on dispersion and densification of cementitious composite comes from their pozolanic reaction and relatevely small particle size (compared to cement grain size) [20, 21].

Although extensive efforts have been made to obtain well dispersed nanotubes in water, it has been found that when incorporated with cementitious composite, a uniform distribution of nanoparticles is may be not found due to the re-agglomeration of CNT. This paper investigates the synergistic effect of usdensified silica fume (a class of microsilica that exhibits extremely fine particle size in a range of 10-200 nm) and MWCNTs on the microsturcure of cementitious composite. Also, this study contributes to a further understanding of the mechanisms of silica particles in the prevention of re-agglomeration of MWCNTs through the composite.

2. Materials and methods

2.1. Materials

The cementitious materials used were ordinary Portland cement (OPC) meeting the requirements of British Standard BS EN 197-1[22], and undensified silica fume 940 U (with ~200 nm mean particle size, ~15m^2/g specific surface area, and >105% 7-day pozzolanic activity index). Polycarboxylate-based superplasticizer (PCE), conforming to ASTM C494 Type F[23], used to impart high workability of the mixture. Silica sand (>99.5 wt.% SiO2, ball milled and sieved to two particle size categories: 0.1–0.18mm and 0.18–0.5 mm). Commercially multiwall carbon nanotube (Cheap Tubes Inc, cat# sku-030102, Brattleboro, VT) with 8–15nm diameter, 10–50 μm length was used in this study. Multiwall carbon nanotubes as provided and undensified Silica fume are shown in Fig.1. Table 1 describes the mix proportions of the two mixes of cement mortar composites in this study.

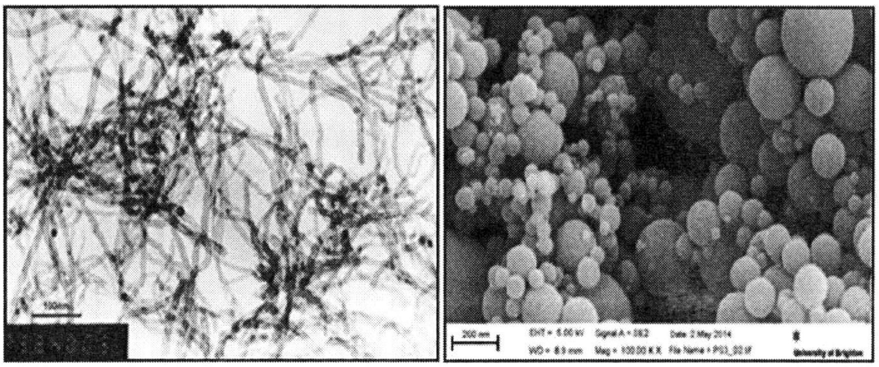

a) b)

Fig. 1: Scanning electron microscope images of a)carbon nanotubes[24],and b) undensified silica fume.

Table 1 Mix proportion of cementitious composite.

Mix	Water/cement ratio	CNTs weight percentage in cement, %	Microsilica USF % of cement	Sand/ cement ratio	SP weight percentage in cement, %
PC_1	0.35	0	0	1.5	0.8
PC_2	0.35	0	0.1	1.5	0.8
CT_1	0.35	0.025	0	1.5	0.8
CT_2	0.35	0.025	0.1	1.5	0.8

2.2. MWCNTs dispersion in mixing water and Sample preparation

A polycarboxylate based superplasticiser (Molecular design of comb-shaped)[25] is an effective dispersant that typically has a charged backbone and uncharged side chains. It has been considered for maintaining composite workability, and can wrap around nanotubes in order to improve their dispersion in water and interaction with cementitious grains. MWCNTs were first dispersed in water in the presence of SP; SP-to-CNF weight ratio of 4.0:1.0 was considered[26].

CNTs were mixed with 25% of the mixing water of cementitious composite, then the suspension was sonicated using a tip ultrasonic probe (Sonic FB-705) set to deliver high intensity ultrasonic energy, which was set to 100% amplitude and operated in 20 s cycles over 5 minutes' duration, following[27].

Four mixes of cementitious martials were prepared for SEM testing: i) plain cement mortar (PC_1), ii) cement mortar with 10% of USF (PC_2), iii) cementitious composite with 0.025% of MWCNTs (CT_1), and iv) cementitious composite with 0.025% of MWCNTs and 10% of USF (CT_2). After dispersion of MWCNTs, dry ingredients including cement, USF, and silica sand were firstly mixed for 2 minutes in the Hobart planetary mixer (Hobart Model A200F) following the ASTM 305[28]. Then water and SP were added (remaining quantity) and after 2 minutes of mixing. The suspension was then added and mixed for a further 4 minutes. Cubes of 50mmx50mmx50mm, and dog-bone shaped briquette in accordance to AASHTO T132 [29] were cast in order to evaluate the compressive and tensile performance of the examined mixes (to be investigated in future work). From the fracture surfaces of cement pastes containing carbon nanotubes, typical samples were taken for microstructure characterisation. MWCNTs and Undensified microsilica characterisation

The morphology of dispersed carbon nanotubes and USF were examined using transmission electron microscopy (TEM) (Hitachi-7100 Transmission Electron Microscope (TEM) operated at 125 kV). To characterize the morphology of MWCNTs, USF, and USF/MWCNTs using TEM, samples of 0.025% MWCNTs, 10% USF (by cement weight), and 10% USF and 0.025MWCNTs were prepared in 100 ml water. Drops of suspension were placed on a grid prior to imaging and allowed to dry in air. Characterisation of the hydrated composites was performed using a high resolution, scanning electron microscope (SEM, Zeiss; model of LEO 1455VP). Prior to SEM imaging fractured surfaces were gold plated and mounted on aluminium stub using copper tape.

3. Results and discussion

TEM images reveal the morphology of the dispersed MWCNT and USF, Fig 2(a, b, and c). The diameter of the individual nanotubes is around 25 nm, and the thickness of the adsorbed layer of SP on the surface of nanotubes is about 10 nm. Spherical particles of USF were found to possess a variety of diameters ranging from 10 nm to more than 200 nm. Fig. 2c (image of CNTs/USF) revealed the synergistic action of silica fume particles in the prevention of re-agglomeration of dispersed nanotubes through the composite, the very small particle size was found in between the neighbouring nanotubes and can act to prevent the re-agglomeration through the hardened phase.

Typical SEM images (Fig. 3 a-d) show the microstructures of cement mortar, and MWCNT/ cementitious composites in the presence and absence of USF. Fig.3 (a and b) represent the effect of USF on the structure of cement paste, USF act to fill the pores system and improve the binding quality of cement hydrate[30]. The small particle size fill up the macropores (more than 50nm) and micrpores(less than 50nm) and leading to refinement the porous structures to a nano scale[31]. (Fig. 3 c-d) revealed that the dispersion states of MWCNTs through the composite was strongly affected by the presence of small particle size of undensified silica fume. Fig. 3a represents composite containing only MWNCTs, although individual nanotubes were observed in suspension through TEM images, a large areas of agglomerated carbon nanotubes through the composites were observed (indicated by the arrows). The re-agglomeration phenomena was attributed to the effect of drying of free water from the mixture during early time of mixture hydration process [32]. When the mixing water (dispersion agent and the media) began to evaporate during cement hydration process, nanotubes move towards each other and form bundles as results of surface tension forces exerted through the evaporation process. During the re-agglomeration process, cement hydration products may not have developed to warp the individual nanotubes before the re-agglomeration of nanotubes occurs.

In case of the presence of USF, the availability of smaller particle size at interfaces between the neighboured nanotubes and cement matrix were effectively hinders the re-agglomeration of dispersed nanotubes. The nanotubes were found to individually embedded through the cement hydration products (Fig. 3b).

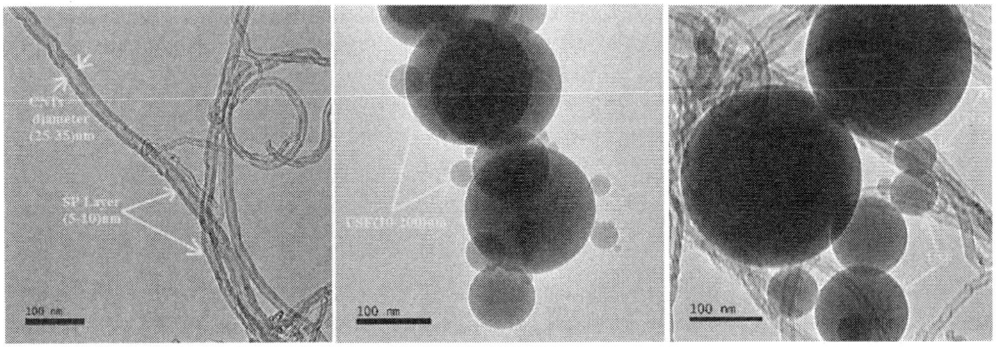

a) b) c)

Fig.2 TEM images showing a) morphology of MWCNTs, b) diameter of USF, and c) the interaction between MWCNTs and MWCNTs.

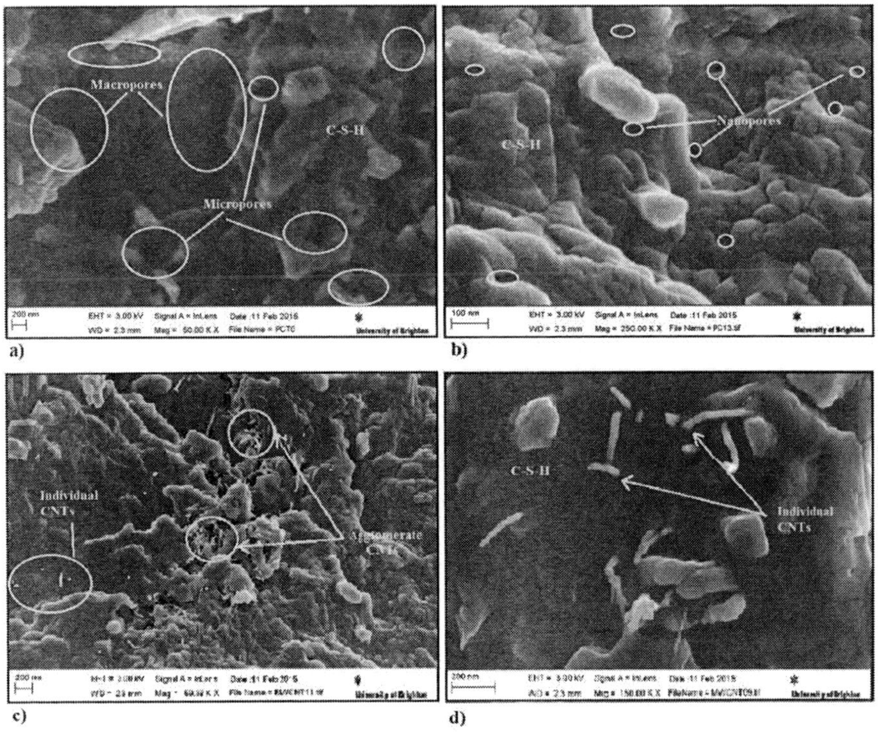

Fig 3 SEM images showing the effect of USF on the structure of plain cement mortar, and the synergistic effect of USF on dispersion of nanotubes in the cementitious matrix a) plain cement

mortar, b) cement mortar with 10% of USF, b) composite without USF, c) composite containing USF and MWCNTs.

4. Conclusion

The effect of USF on the dispersion of MWCNTs through the cementitious composite has been studied. The microstructure analysis revealed that USF particles were found to facilitate the dispersion of CNTs and improve the interfacial interaction between carbon nanotubes and the surrounding matrix. These can be attributed to the pozzolanic reaction and small particle size relative to the size of cement grain. The observed small particle size of silica fume (in range from 10 nm) was found to act as a physical barrier reducing the possibility of nanotube re-agglomeration within the cementitious composite. This should lead to an improvement in the physical performance of the composite.

Acknowledgments
The lead author would like to thank the Iraqi Ministry of Higher Education and Scientific Research, Iraqi Culture attaché in London, Babylon University in Iraq for award of a Doctoral scholarship

5. Reference

1. Iijima, S., *Carbon nanotubes: past, present, and future*. Physica B: Condensed Matter, 2002. **323**(1-4): p. 1-5.
2. Ma, P.-C., et al., *Dispersion and functionalization of carbon nanotubes for polymer-based nanocomposites: A review*. Composites Part A: Applied Science and Manufacturing, 2010. **41**(10): p. 1345-1367.
3. Konsta-Gdoutos, M.S., Z.S. Metaxa, and S.P. Shah, *Multi-scale mechanical and fracture characteristics and early-age strain capacity of high performance carbon nanotube/cement nanocomposites*. Cement and Concrete Composites, 2010. **32**(2): p. 110-115.
4. Z. S. Metaxa, M.S.K.-G., and S. P. Shah, *Mechanical Properties and Nanostructure of Cement-Based Materials Reinforced with Carbon Nanofibers and Polyvinyl Alcohol (PVA) Microfiber*. 2009.
5. Jose , L.F.J.M.d.C., and Juan Ángel García, *Carbon Nanotube-Cement Composites in the Construction Industry: 1952-2014. A State of the Art Review*. 2nd International Conference on Emerging Trends in Engineering and Technology (ICETET'2014), 2014.
6. Yu M, L.O., Dyer MJ, Kelly TF, Ruoff RS., *Strength and breaking mechanism of multiwalled carbon anotubes under tensile load. Science 2000;287:637–40*. 2000.
7. Wong EW, S.P., Lieber CM., *Nanobeam mechanics: elasticity, strength, and toughness of nanorods and nanotubes*. . 1997. **277:1971–5.**
8. Kharissova, O.V., B.I. Kharisov, and E.G. de Casas Ortiz, *Dispersion of carbon nanotubes in water and non-aqueous solvents*. RSC Advances, 2013. **3**(47): p. 24812.
9. Dong, B., et al., *Dispersion of carbon nanotubes by carbazole-tailed amphiphilic imidazolium ionic liquids in aqueous solutions*. J Colloid Interface Sci, 2011. **356**(1): p. 190-5.
10. Sobolkina, A., et al., *Dispersion of carbon nanotubes and its influence on the mechanical properties of the cement matrix*. Cement and Concrete Composites, 2012. **34**(10): p. 1104-1113.
11. Li, G.Y.W., Pei Ming Zhao, Xiaohua, *Mechanical behavior and microstructure of cement composites incorporating surface-treated multi-walled carbon nanotubes*. Carbon, 2005. **43**(6): p. 1239-1245.

12. Trettin R, K.T., *Nanotubes fur hochleistungsbetone (nanotubes for highperformance concretes)*. Betonwerk und Fertigteil-Technik/Concr Precast Plant Technol, 2005. **71**(2): p. 20–1.
13. Musso, S., et al., *Influence of carbon nanotubes structure on the mechanical behavior of cement composites*. Composites Science and Technology, 2009. **69**(11-12): p. 1985-1990.
14. Jyoti Bharj, S.S., Subhash Chander, Rabinder Singh, *Role of Dispersion of Multiwalled Carbon Nanotubes on Compressive Strength of Cement Paste*. World Academy of Science, Engineering and Technology International Journal of Physical, Nuclear Science and Engineering, 2014. **Vol:8 No:2**.
15. Li, H., H.-g. Xiao, and J.-p. Ou, *A study on mechanical and pressure-sensitive properties of cement mortar with nanophase materials*. Cement and Concrete Research, 2004. **34**(3): p. 435-438.
16. Chaipanich, A.N., Thanongsak. Wongkeo, Watcharapong. Torkittikul, Pincha, *Compressive strength and microstructure of carbon nanotubes–fly ash cement composites*. Materials Science and Engineering: A, 2010. **527**(4–5): p. 1063-1067.
17. Morsy, M.S., S.H. Alsayed, and M. Aqel, *Hybrid effect of carbon nanotube and nano-clay on physico-mechanical properties of cement mortar*. Construction and Building Materials, 2011. **25**(1): p. 145-149.
18. Sanchez, F.I., Chantal, *Microstructure and macroscopic properties of hybrid carbon nanofiber/silica fume cement composites*. Composites Science and Technology, 2009. **69**(7–8): p. 1310-1318.
19. Setina, J., A. Gabrene, and I. Juhnevica, *Effect of Pozzolanic Additives on Structure and Chemical Durability of Concrete*. Procedia Engineering, 2013. **57**: p. 1005-1012.
20. Kim, H.K., I.W. Nam, and H.K. Lee, *Enhanced effect of carbon nanotube on mechanical and electrical properties of cement composites by incorporation of silica fume*. Composite Structures, 2014. **107**: p. 60-69.
21. Sanchez, F. and A. Borwankar, *Multi-scale performance of carbon microfiber reinforced cement-based composites exposed to a decalcifying environment*. Materials Science and Engineering: A, 2010. **527**(13–14): p. 3151-3158.
22. *BS EN 197-1:2000*, in *Cement. Composition, specifications and conformity criteria for common cements*.
23. *ASTM C494 / C494M-16, Standard Specification for Chemical Admixtures for Concrete, ASTM International, West Conshohocken, PA, 2016, www.astm.org*.
24. Inc, C.T. *https://www.cheaptubes.com/product/multi-walled-carbon-nanotubes-8-15nm/*. 2016.
25. Winnefeld, F., et al., *Effects of the molecular architecture of comb-shaped superplasticizers on their performance in cementitious systems*. Cement and Concrete Composites, 2007. **29**(4): p. 251-262.
26. Konsta-Gdoutos, M.S., Z.S. Metaxa, and S.P. Shah, *Highly dispersed carbon nanotube reinforced cement based materials*. Cement and Concrete Research, 2010. **40**(7): p. 1052-1059.
27. S. Alrekabi, A.C., A. Lampropoulos, I. Savina, *Experimental Investigation on the Effect of Ultrasonication on Dispersion and Mechanical Performance of Multi-Wall Carbon Nanotube-Cement Mortar Composites*. International Journal of Civil, Environmental, Structural, Construction and Architectural Engineering, 2016. **10**: p. 267-274.
28. *ASTM C305-14, Standard Practice for Mechanical Mixing of Hydraulic Cement Pastes and Mortars of Plastic Consistency, ASTM International, West Conshohocken, PA, 2014, www.astm.org*.
29. *AASHTO T 132-87 Standard Method of Test for Tensile Strength of Hydraulic Cement Mortars, STANDARD by American Association of State and Highway Transportation Officials, 1987*. 2013.

30. Ahmed Sbia, L., et al., *Enhancement of Ultrahigh Performance Concrete Material Properties with Carbon Nanofiber*. Advances in Civil Engineering, 2014. **2014**: p. 1-10.
31. Yu Hu, D.L., Penghui Li, Qingbin Li, Guoqiang Sun, *Fracture toughness enhancement of cement paste with multi-walled carbon nanotubes*. Construction and Building Materials, 2014. **70 (2014) 332–338**.
32. Mendoza, O., G. Sierra, and J.I. Tobón, *Effect of the reagglomeration process of multi-walled carbon nanotubes dispersions on the early activity of nanosilica in cement composites*. Construction and Building Materials, 2014. **54**(0): p. 550-557.

IOP Conf. Series: Journal of Physics: Conf. Series **829** (2017) 012012 doi:10.1088/1742-6596/829/1/012012

Efficient and rapid degradation of Congo red dye with TiO_2 based nano-photocatalysts

Himanshu Narayan and Hailemichael Alemu

Faculty of Science & Technology, National University of Lesotho, Roma 180, Lesotho

h.narayan@nul.ls

Abstract. Degradation of Congo red (CR) dye with TiO_2 based nano-photocatalyst (NPC) loaded with Nd^{3+} and Er^{3+} ions is reported. The chemical route of synthesis through co-precipitation/hydrolysis (CPH) was employed to produce NPCs with general composition $TiO_2[R_2O_3]_x$, $\{x = 0.1, 0.2; R \equiv Nd, Er\}$ and particle size within 12 – 16 nm. Photocatalytic degradation under visible light was measured in terms of the percent degradation of CR in 180 min (C'_{180}), time taken to degrade to half of the initial CR concentration ($t_{1/2}$) and apparent rate constant (k_{obs}). For both doping types, values of C'_{180} close to 100% were obtained with $x = 0.2$ NPCs, indicating complete removal of the dye. For the same NPCs, very high values of k_{obs} were found; 2.91×10^{-2} min^{-1} and 2.36×10^{-2} min^{-1}, for Nd^{3+} and Er^{3+} loaded NPCs, respectively, suggesting very rapid degradation. Other NPCs with $x = 0.1$, also showed reasonably good and fast degradation of CR. The observations may be attributed to the small particle size of the NPCs. Moreover, from the DRS results it is observed that the addition of Nd^{3+} and Er^{3+} ions apparently introduces intermediate energy levels within the band gap of TiO_2. Such new levels seem to support photocatalysis because they act as electron traps leading to effective suppression of the undesired e^-/h^+ recombination. To some meaningful extent they also facilitate the absorption of visible irradiations required in the process.

1. Introduction

Heterogeneous photocatalysis has been an attractive method since 1970s to treat several pollutants. Titanium oxide (TiO_2) has been identified as one of the most effective photocatalyst for the purpose of degradation of pollutants from wastewater. However, it requires expensive UV irradiation for photocatalysis to take place. Utilization of visible light in TiO_2 based photocatalysis can make the process of treatment more cost-effective.

Photocatalysis can involve two fundamental processes: photo-excitation and photosensitization. In the first process the electrons from valence band (VB) need to be excited and promoted to the conduction band (CB), thus producing pairs of negative-electron and positive-holes (e^-/h^+ pairs). Here the energy-gap plays a vital role because it determines the energy required for the process to start. In the second process, an adsorbed pollutant (e.g., dye) molecule gets photo-excited initially leading to an interaction with the catalyst. As a result, electrons of adsorbed molecule are subsequently transferred to the catalyst particle. In this process, the particle-size becomes important because smaller size can offer larger specific area for adsorption. Nevertheless, in both processes the creation of e^-/h^+ pairs is critical, because these particles individually participate in the production of highly active species that eventually destruct the pollutants chemically. Due to the same reason, the recombination of e^-/h^+ pairs, which depletes significant portion of the electrons and holes produced, needs to be controlled for better photocatalytic efficiency [1-3].

In order to make photocatalysis possible with pure TiO_2 under visible light, the 3.2 eV band gap (i.e., cutoff wavelength, $\lambda_{cutoff} \sim 390$ nm) needs to be narrowed down. On the other hand, specific surface area of the catalyst, to facilitate enhanced degradation, can be increased through synthesis of smaller particles. However, it should be noted that particles with less than about 10 nm size may impede the degradation, because at this size, (i) the detrimental e^-/h^+ recombination may become substantially high owing to smaller mean-free-path, and (ii) the band gaps may become wider due to quantum confinement. Both of these effects would result in undesirable blue-shift of λ_{cutoff} away from the visible regions [2,3]. We have found that the most appropriate range of particle size is 10 – 20 nm [2,3].

Our recent investigations have shown that narrowing of band-gap may be of lesser importance than the reduced particle size of the photocatalyst. Moreover, effective control of the unwanted e^-/h^+ recombination through appropriate doping or mixing may be the key to additional improvement of degradation. In this direction, we have already noted that rare-earth ions could be the most suitable choice as dopants [2] because of their ability to efficiently suppress e^-/h^+ recombination. Further, because to their unique electronic structure, the rare-earths can possibly modify the band-gap of TiO_2 also, such that visible light photocatalysis through photo-excitation is supported. Along these lines, we report in this paper about the photocatalytic activity of TiO_2 based nanocomposites loaded with Nd^{3+} and Er^{3+} ions. Some of these materials have produced very rapid degradation of the dye Congo red (CR) with close to 100% removal of the dye.

2. Experimental

2.1. Materials synthesis

Titanium oxide nanocomposites loaded with Nd^{3+} and Er^{3+} ions, i.e., with general formula $TiO_2[R_2O_3]_x$ $\{x = 0.1, 0.2; R \equiv Nd, Er\}$ were synthesized through the method of co-precipitation/hydrolysis (CPH). The details about sample preparation are reported elsewhere [2,3]. Chemicals used in the process of synthesis were neodymium nitrate (99.9% metal basis), erbium nitrate (99.9% metal basis), isopropanol and titanium(IV)-butoxide [$Ti(OBu)_4$] (97% reagent grade), supplied by Aldrich Chemical Corporations. Ammonium hydroxide was obtained from Shalom Laboratory Supplies and Congo red was procured from E. Merck Co. (Darmstadt, Germany).

2.2. Measurements

Powder X-ray diffraction (XRD) patterns were recorded with Shimadzu D6000 diffractometer (from Shimadzu, Japan) using CuKα radiation (λ = 1.5406 Å) and transmission electron microscopy (TEM) was carried out on a Philips CM200 TEM instrument (from FEI Co.). Fourier transform infra-red spectroscopy (FTIR) and Raman spectroscopy were carried out on Jasco FTIR 4100 set up (from Jasco) and on inVia Raman microscope (from Renishaw) units, respectively. A source wavelength of 514 nm was used in Raman analysis. The diffused reflectance spectroscopy (DRS) was carried out to record the absorption spectra of the samples using a Varian CARY 5000 UV-VIS-NIR absorption spectrometer (from TNLabs).

Visible-light photocatalytic degradation of CR was carried out on a locally fabricated cost-effective photoreactor. A series of control experiments were performed prior to the actual investigations in order to observe separately the effects of both the photocatalyst and the visible light. First, the absorbance of an experimental solution without any NPC under light irradiation, and then in the dark but with a solution that contained the NPC, were recorded. It was noticed that the catalyst and light alone could not produce any meaningful degradation. Apart from this, the amounts of photocatalyst

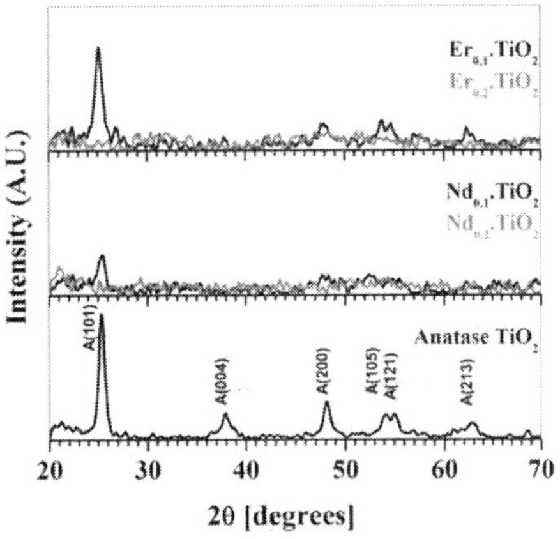

Figure 1: XRD plots of the Nd^{3+} and Er^{3+} loaded TiO_2 NPC samples.

and CR in the experimental solution were also optimized for best results. The step-by-step details about these measurements are already reported elsewhere [3].

Results and Discussion

The XRD results for the NPC samples of $TiO_2[R_2O_3]_x$, $\{x = 0.1, 0.2; R \equiv Nd, Er\}$ show that most of

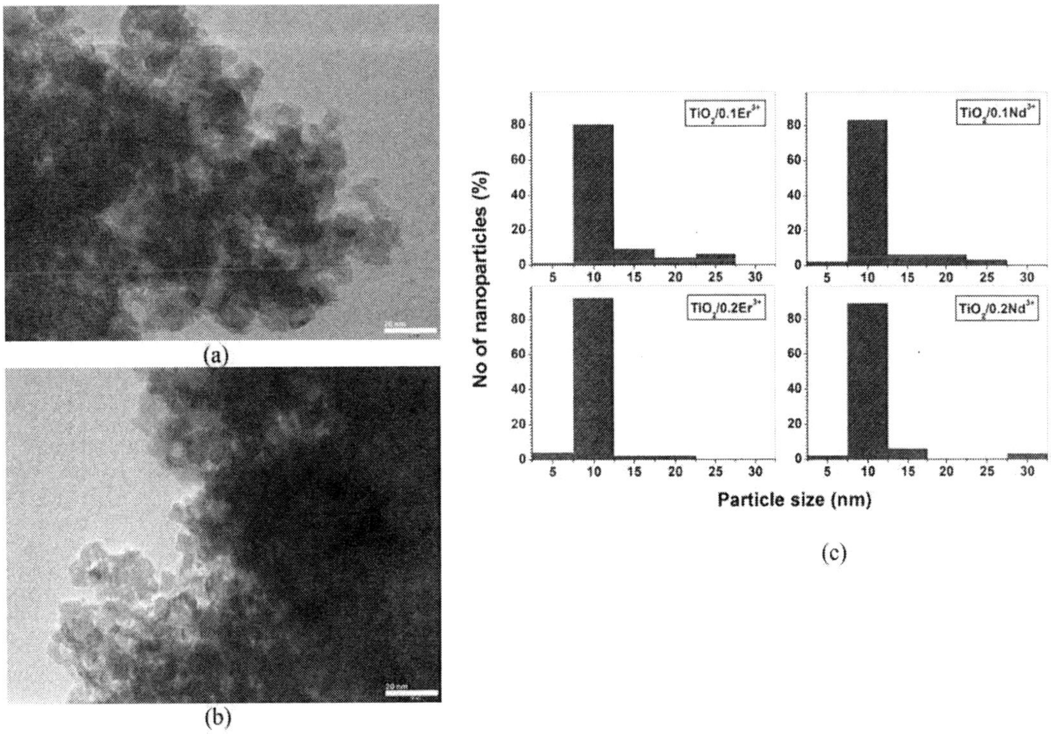

Figure 2: Typical TEM images of the NPCs: (a) Nd^{3+} and (b) Er^{3+} loaded TiO_2 samples (The scale shown is equivalent to 20 nm), and (c) particle-size distribution of all samples.

the peaks are highly broadened [Figure 1], because of the smaller crystallite size of the samples. Nevertheless, the dominating TiO_2 phase in all the samples, is confirmed by the presence of the most prominent peak around $2\theta = 25.2°$, which corresponds to the (101) reflection of anatase. Compared to the XRD data for pure anatase however, the slightly shifted peaks of NPC indicate that the rare-earth ions probably occupy the lattice sites. Further, the presence of cubic Er_2O_3 and hexagonal Nd_2O_3 phase in the composites is also confirmed. The XRD results also established that overall homogeneity of the composites was maintained and no new phase was formed.

The XRD data was also used to roughly estimate the particle- (crystallite-) size z, through Debye-Scherrer equation [2,3];

$$z = \frac{k\lambda}{\beta \cos\theta},$$

where, the shape-factor k was taken as 0.9, assuming spherically symmetric crystallites in the NPCs. For the Nd^{3+} loaded samples, estimated particle size were around 13 and 15 nm, respectively, for x = 0.1 and 0.2. The same were estimated for x = 0.1 and 0.2, Er^{3+} loaded samples, as 16 and 12 nm, respectively.

Figure 2 (a) and (b) depict the TEM images of the NPCs [shown only for x = 0.1 loaded samples], which facilitated the analysis of the crystallinity and morphology of the samples. Mostly spherical and ellipsoidal particles seem to agglomerate into bigger particles forming bulk of the NPCs. Particle size was also measured directly from TEM images using the ImageJ software. Typical distributions of particle size obtained from these pictures are shown in figure 2 (c). Clearly, majority of the NPC particles, i.e., close to 80% or more, have the size within 10 – 15 nm. The size varies from around 8 to 16 nm for Nd^{3+} loaded, and from around 9 to 17 nm for Er^{3+} loaded samples. The mean particle size of around 12 nm and 14 nm for both x = 0.1 NPCs was measured respectively for Nd^{3+} and Er^{3+} loaded samples. Similarly, for the x = 0.2 doped samples, the average size were found to be nearly 15 and 12 nm, respectively. These values were in good agreement with the size estimated from XRD data. All the estimated and measured particle size are listed in Table 1.

Figures 3 and 4 show the typical Raman and FTIR spectra for the NPCs, respectively. In general, five characteristic peaks around 144, 197, 399, 513 and 639 cm^{-1} are clearly visible in the Raman

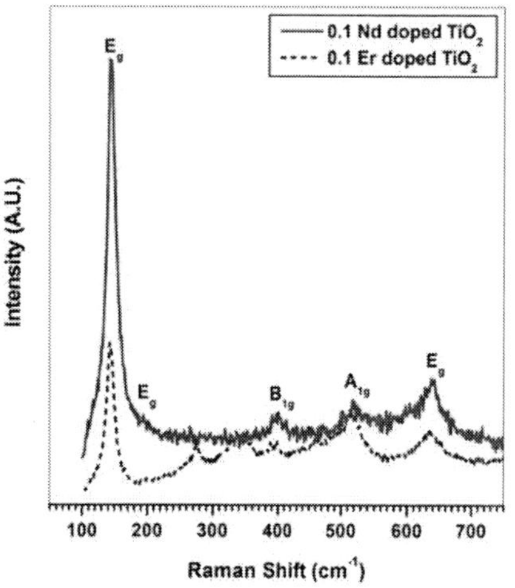

Figure 3: Raman spectra of Nd^{3+} and Er^{3+} loaded TiO_2 samples (x = 0.1).

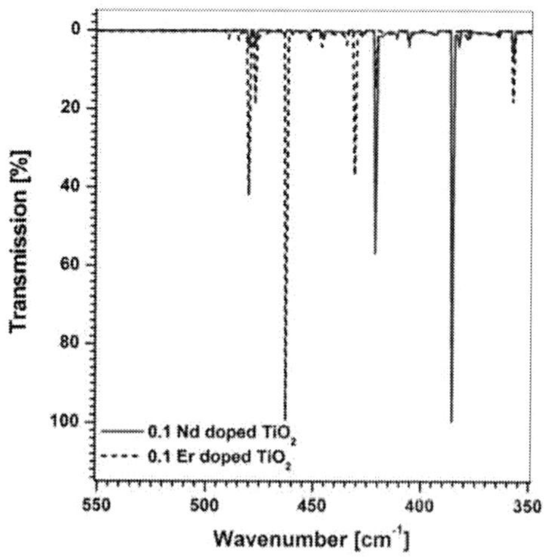

Figure 4: FTIR spectra of Nd^{3+} and Er^{3+} loaded TiO_2 samples (x = 0.1).

profiles obtained with our samples, which may be associated with the anatase TiO_2. Usually, the presence of anatase is identified through five distinct Raman peaks at about 151, 196, 409, 515 and 633 cm^{-1}, assigned to the E_g, E_g, B_{1g}, A_{1g} or B_{1g} and E_g modes, respectively [4,5]. However, the peaks corresponding to E_g and B_{1g} modes in our Raman spectra are apparently red-shifted to around 144 and 399 cm^{-1}, respectively. Such red-shifts may be explained on the basis of a phonon confinement model, and are associated with reduced particle size of the material under consideration [2,4]. This observation is in agreement with the XRD results for all the NPCs. Further, Raman spectroscopy results also indicate suppression of the Ti–O bonds due to loading of rare-earth ions. The FTIR spectra (Figure 4) of pure TiO_2, in general, shows multiple peaks in the region 420 cm^{-1} to 550 cm^{-1}, which are usually associated with the various characteristic vibrational modes of Ti–O bonds [5]. For loaded NPC samples however, the intense peaks at wavenumber 492 cm^{-1} and 434 cm^{-1} are missing and overall peak intensity is reduced in the FTIR spectra, which indicates that the dopant ions have suppressed the Ti–O bonds. This observation was supported also by Raman spectroscopy results as mentioned earlier. The absence of characteristic vibration modes for Nd and Er and their oxides Nd_2O_3 and Er_2O_3 in the Raman and FTIR spectra suggests that there was no segregation of these elements or the oxides in TiO_2 matrix. It confirms the homogeneity of the NPC samples. Moreover, both the Raman and FTIR results do not show any new peak corresponding to either R–O or Ti–O bonds.

The UV-vis absorption data obtained from diffused reflectance spectroscopy (DRS), plotted in figure 5 (a), were used to estimate the band-gaps of all the NPC samples using the Tauc's equation, $\alpha h\nu = A\left(h\nu - E_g\right)^n$, with $n = 1/2$, for direct allowed transitions [6]. Figure 5 (b) shows the plots of $\left(\alpha h\nu\right)^2$ against $h\nu$. From these plots, the intersection of the tangent to the most linear part of the curve, with energy axis gives the value of energy-gap. Interestingly, the band-gap for all the NPCs loaded with Nd^{3+} and Er^{3+} ions was determined to be 3.35 eV. On the other hand, the band-gap for pure anatase (not shown), following the same method, was estimated to be 3.2 eV. Clearly, the loading of TiO_2 with rare-earth ions seems to widen the band-gap, which is opposite to what was initially desired. Widening of band-gap is expected to push the absorption-edge, λ_{cutoff} more towards the shorter

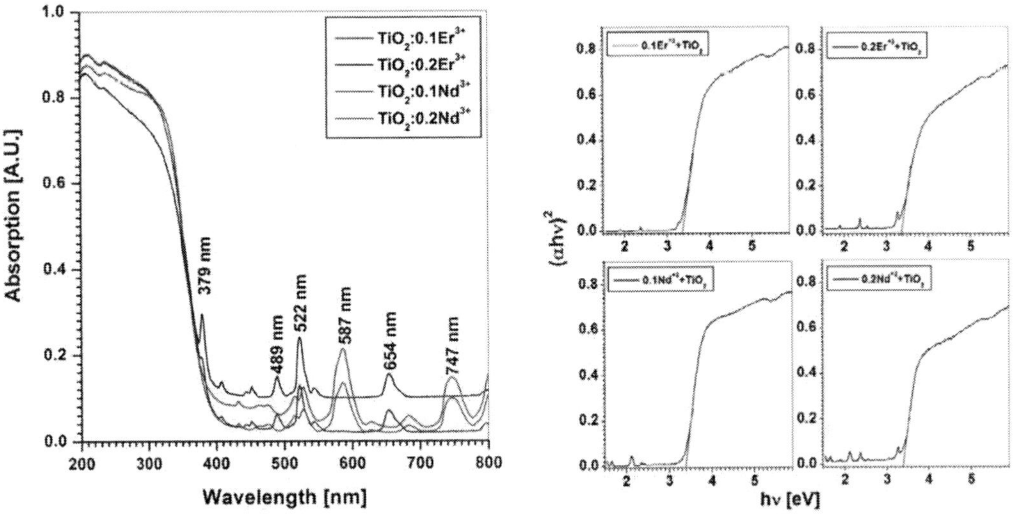

(a) (b)

Figure 5: DRS spectra (a) and the Tauc's plots obtained from DRS spectra (b) for the TiO_2 NPCs loaded with Nd^{3+} and Er^{3+}.

SUMMARY OF RESULTS

Sample	x	Particle-size	k_{obs} *	$t_{1/2}$	C'_{180}
		nm	*10^{-2} min^{-1}*	*min*	%
CPH $TiO_2.[Nd_2O_3]_x$	0.1	13	2.01 (3.58)	18	97
	0.2	15	2.91 (2.91)	16	100
CPH $TiO_2.[Er_2O_3]_x$	0.1	16	1.03 (1.70)	20	90
	0.2	12	2.36 (3.88)	11	99

* The quantities in the brackets in k_{obs} column are the initial values for apparent rate constants calculated using the first four points (i.e., up to 90 min of degradation).

wavelengths (UV region) instead of bringing it inside the visible range to facilitate photocatalysis under visible light. However, a careful examination of the original absorption (DRS) spectra [figure 5 (a)] shows some absorption peaks in the visible regions. Among these, the peak at 587 nm for Nd^{3+} doped NCs, and the one at 522 nm for Er^{3+} doped NCs cannot be ignored. The 528 and 587 nm absorption peaks have already been reported and their appearance has been attributed to the formation of intermediate energy levels between the VB and CB of TiO_2 due to the 4*f* shells of Nd^{3+} ions [7-9]. In the same line, the absorption peaks at 489, 522 and 654 nm observed with Er^{3+} doped NCs may also be associated with the formation of similar energy levels in these composites. Expectedly, these new energy levels in the band-gap of TiO_2 should facilitate both the suppression of e^-/h^+ recombination though trapping of electrons, and absorption of visible radiations through effective narrowing of band-gap.

Figure 6 shows the degradation of CR in terms of the relative percentage [$(C/C_0)\times 100$] with time up to 180 min. One can immediately observe from this figure that in less than about 25 min, all the NPCs degrade the dye to half of the starting concentration or even more. It clearly indicates that the NPCs are very fast catalysts. In comparison, the as-received anatase powder (particle size around 57 nm) seems to take more than three hours to reach maximum achievable degradation of CR under the same experimental conditions.

For all the NPC samples, the rate of degradation was found to be of first order. From a least-square regression of $\ln(C/C_0)$ vs time data, the apparent rate constant (k_{obs}) was estimated. The best

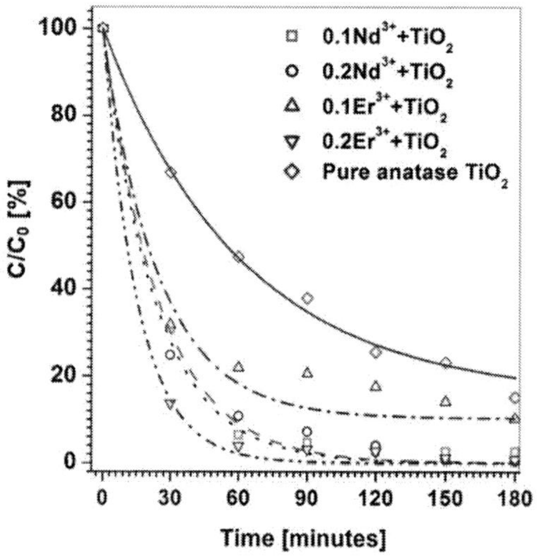

Figure 6: Photocatalytic degradation of CR with TiO_2 NPCs loaded with Nd^{3+} and Er^{3+}.

performing sample in terms of degradation after 180 min (C'_{180}) was Nd^{3+} ($x = 0.2$) loaded TiO_2 NPC, which degraded close to 100% of the dye ($C'_{180} = 100\%$, $k_{obs} = 2.91 \times 10^{-2}$ min^{-1}). This performance was only marginally better than the Er^{3+} ($x = 0.2$) loaded TiO_2 NPC for which $C'_{180} = 99\%$ at $k_{obs} = 2.36 \times 10^{-2}$ min^{-1}. The initial rate of degradation (calculated using the first four points only in the graph) however, for this NPC was much higher at 3.88×10^{-2} min^{-1}. Because of such high initial rate, this NPC degraded more than half of the initial CR concentration in just about 11 min ($t_{1/2}$), and nearly 98% in just one hour. For the $x = 0.1$ Nd^{3+} and Er^{3+} loaded NPCs, C'_{180} was found to be 97 and 90%, respectively. For the pure anatase powder, on the other hand, C'_{180} was only about 80%. The relatively poor performance of the as-received anatase powder may be directly related to the larger average particle size of about 57 nm. However, for the NPCs loaded with Nd^{3+} and Er^{3+} ions, not just the smaller size, but a number of other factors too, contribute to the better degradation efficiency. The importance of loading of rare-earth ions and its exclusive contribution to degradation has been critically discussed in some of our earlier reports [2,3].

A combination of some crucial intrinsic properties of the composites determines the final degradation results. Most important among these are the particle size and the ability to control e^-/h^+ recombination. Apart from these two, the sensitivity of the NPCs to visible light also plays a significant role if the loading modifies the band-gap for absorption of the latter. In the $x = 0.2$ Er^{3+} loaded NC, $C'_{180} = 99\%$ may be due to the smaller average particle size of 12 nm. The $x = 0.1$ Er^{3+} loaded NPC, on the other hand, produced $C'_{180} = 90\%$, apparently because it had a marginally bigger average particle size of 14 nm. Besides the smaller particle size of the NPC, the loading of Nd^{3+} and Er^{3+} ions was also critical because: (i) it helped efficient suppression of e^-/h^+ recombination; and (ii) it introduced intermediate energy states in the energy gap of the host TiO_2, which may have enhanced the visible light sensitivity of the composites. Intermediate levels may be introduced in semiconductors through doping of rare-earths [7-9], which is confirmed from our DRS results as discussed earlier. These intermediate, rare-earth related energy levels may not only act as electron-traps, but they could also facilitate the absorption of visible radiations through indirect narrowing of the band-gap. Upon irradiation with visible light, the electrons produced in the molecules (pollutant) adsorbed on the TiO_2 particles can be captured on these traps, and thus the holes would be left alone to freely participate in the chemical processes leading to degradation (suppression of e^-/h^+ recombination). Moreover, under the changed scenario, the electrons in the VB of TiO_2 would require less excitation energy, which could possibly be supplied from visible light irradiation, to get transferred to the rare-earth related intermediate energy levels. Thus, the charge-separation necessary for the process may be supported and some additional photocatalytic degradation through photo-excitation may also take place.

Conclusions

We have reported the visible light photocatalytic activity of TiO_2 base nano-photo-composites loaded with Nd^{3+} and Er^{3+} ions; $TiO_2[R_2O_3]_x$, $\{x = 0.1, 0.2; R \equiv Nd, Er\}$, in the degradation of Congo red dye. Close to 100% degradation at a very rapid rate has been produced with both types of NPCs having $x = 0.2$. The other two compositions of NPCs with $x = 0.1$, also showed reasonably significant degradation of CR. The observations have been attributed to the smaller particle size (12 – 16 nm) of the NPCs. It has been concluded that doping of rare-earth ions further supports photocatalysis through introducing intermediate energy levels in the energy gap of host TiO_2, which may act like efficient electron traps for suppression of the unwanted e^-/h^+ recombination. At the same time, these energy levels may also narrow down the energy gap indirectly, so that visible light induced photocatalysis through photo-excitation also becomes possible to some extent.

Acknowledgments

We would like to thank Professor Madhavi Thakurdesai and Ms Smita Survase of Thin Film Research Laboratory, Department of Physics, Birla College (Affiliated to University of Mumbai), Kalyan, India for their help in some of the characterization experiments.

References

[1] K. Hashimoto, H. Irie, and A. Fujishima, "TiO2 photocatalysis: A historical overview and future prospects," Jap. J. Appl. Phys., vol. 44 (12), pp. 8269-8285, 2005.

[2] H. Narayan, H. Alemu, L. Macheli, M. Thakurdesai, and T. K. Gundu Rao, "Synthesis and characterization of Y^{3+}-doped TiO_2 nanocomposites for photocatalytic applications," Nanotechnology, vol. 20, 255601 (8 pp), 2009.

[3] H. Narayan, H. Alemu, L. Macheli, M. Sekota, M. Thakurdesai, and T. K. Gundu Rao, "Role of particle size in visible light photocatalysis of Congo red using $TiO_2[ZnFe_2O_4]_x$ nanocomposites," Bull. Mater. Sci., vol. 32 (5), pp.499-506, 2009.

[4] W. F. Zhang, Y. L. He, M. S. Zhang, Z. Yin, Q. Chen, "Raman scattering study on anatase TiO_2 nanocrystals," J. Phys. D: Appl. Phys., vol. 33, pp. 912-916, 2000.

[5] M. J. Šćepanović, M. Grujić-Brojčin, Z. D. Dohčević-Mitrović, Z. V. Popović, "Characterization of anatase TiO_2 nanopowder by variable-temperature Raman spectroscopy," Sci. Sintering, vol. 41, pp. 67-73, 2009.

[6] Anil Tumuluri, K. Lakshun Naidu, K. C. James Raju, "Band gap determination using Tauc's plot for LiNbO3thin films," Int. J. Chem. Tech. Res., vol. 6 (6), pp. 3353-3356, 2014.

[7] X. Su, J. Zhao, Y. Li, Y. Zhu, X. Ma, F. Sunb, and Z. Wang, "SolutionSynthesis of Cu_2O/TiO_2 Core-Shell nanocomposites," Colloids Surf.-A, vol. 349, 151-155, 2009.

[8] W. F. Zhang, Y. L. He, M. S. Zhang, Z. Yin, and Q. Chen, "Raman scattering study on anatase TiO_2 nanocrystals," J. Phys. D: Appl. Phys., vol. 33, pp. 912-916, 2000.

[9] M. A. J. Klik, T. Gregorkiewicz, I. V. Bradley, and J. P. R. Wells, "Optically induced deexcitation of rare-earth ions in a semiconductor matrix," Phys. Rev. Lett., vol. 89 (22), 227401 (4 pp), 2002.

IOP Conf. Series: Journal of Physics: Conf. Series **829** (2017) 012013 doi:10.1088/1742-6596/829/1/012013

Role of hydrogen in affecting the growth trend of CNT on micron spherical silica gel

R N Othman[1,*], A I Ismadi[1], and S N Tawil[2]

[1]Department of Mechanical Engineering, National Defence University Malaysia, Kem Sungai Besi, 57000, Kuala Lumpur, Malaysia.

[2]Department of Electrical and Electronics Engineering, National Defence University Malaysia, Kem Sungai Besi, 57000, Kuala Lumpur, Malaysia.

*Corresponding author: izawati@upnm.edu.my

Abstract. Grafting CNTs onto substrates such as fibres and microparticles offers an alternative approach to tackle the issues associated with dispersion in a composite matrix, as well as additional benefits (hybrid effects) provided by these dual-filler systems. One approach to obtain such hybrid systems is the direct growth of nanotubes on the supporting fibre or particles. Previous study has shown that the CNTs would grow on the silica microparticles with the morphology closely related to the operating conditions such as temperature and time. However, the role of hydrogen in affecting the tube's morphology was not explored before. The particles were synthesized via chemical vapour deposition (CVD) method. Spherical silica gel with 40 - 75 μm diameter was used as the substrate. Toluene and ferrocene were used as the hydrocarbon and catalyst source, respectively. The reaction time was kept for four hours while the temperature was maintained at 850°C. The FESEM and TEM investigation proved that the flow hydrogen during reaction caused a tremendous difference in the outer diameter of the synthesized CNTs. Relatively thin CNT was observed under 50 ml/min of hydrogen flow compared to the particles synthesized without hydrogen. Raman spectroscopy of the CNTs revealed three bands; the disorder-induced D mode (~1321 cm^{-1}), the tangential G mode (~1570 cm-1) and second order G' mode (~2642 cm^{-1}). Raman analysis shows that the synthesized CNTs exhibited all these peaks, confirming the existence of CNTs. As G peak is more intense than D peak for all samples synthesized under hydrogen flow, it can be concluded that CNTs synthesized is indeed of high quality. It can be confirmed that hydrogen plays an important role in influencing the morphology of the synthesized tubes.

1. Introduction

Carbon Nanotube (CNT) gained its popularity as material with extraordinary mechanical, electrical, and thermal properties. The applications that utilized the CNTs cover a wide range; which include as filler in composites, in vivo biomedical application, solar thermal applications; just to name a few. As such, the dimensions and morphology of the CNTs need to be precisely controlled. However, CNTs tend to agglomerate, which causes a real problem if they were to be used as filler in composites. Various work reported on tackling the dispersion issues such as applying high shear force [1], intensive tip sonication [2] and/or mixing the CNTs with surfactant [3]. These processes however involve many steps that proven difficult for industrial processing, which may results in the change in the overall composite processing cycle [4].

Growing CNTs on a substrate via chemical vapor deposition method (CVD) offers an alternative in introducing the CNTs into polymer matrix [5]. The examples of the substrates used are carbon fiber [6], silica fiber [7], and alumina oxide [8]. These CNTs were grown directly on the substrates and become strongly attached to them. As the substrate is normally in micron size, the agglomeration and within the polymer matrix is hindered. Ultimately, the viscosity increment of the resulting suspension became minimum [9].

Several reports have mentioned about the CVD parameters involved during producing these hybrid structures [10]. However, no detail report has been focused on the effects of varying hydrogen flow rate on this porous spherical substrate. Previous study reported the growth of CNT on porous substrate [5] but without the hydrogen flow rate. There was also a report on the effects of hydrogen flow rate where it was reported that hydrogen affects the CNT morphology and crystalline structure [11]. Hydrogen was also reported to etch pyrolytic carbon from the catalyst which eventually delaying the catalyst poisoning [12]. This study investigates the effects of co-flowing hydrogen into CVD on the morphology and crystallinity of the CNTs obtained as well as the amount of CNTs obtained.

2. Experimental

In this work, the CNT was produced under atmospheric pressure in a horizontal CVD. The set up of the CVD system is similar to the one published previously [5], which consists of an electrical furnace, quartz tube, liquid feeding system, and pre heater. Prior to that, the temperature profile along the furnace was performed to identify the area, which constitutes the temperature closest to the set temperature. The zone was located within 10 – 20 cm from the entrance. As such, the substrate and subsequently the CNT product will be collected from this zone.

The synthesis of CNT on spherical silica gel was performed in this CVD system by introducing catalyst via non-floating method. Silica gel was purchased from Sigma Aldrich, which is of high purity grade. According to the safety date sheet, the pore diameter of the porous silica was reported to be 6 nm and the size of the gels in the range between 40 to 75 μm. Toluene and Ferrocene (both purchased from Sigma Aldrich) were used as hydrocarbon source and catalyst, respectively. First, the quartz boat containing silica and catalyst (1:1 wt. %) was placed within the heating zone. Then, the temperature was increased to 850°C while flowing the Argon gas at 300 ml/min. The furnace was left for 10 minutes after the temperature reached 850°C, to allow thermal equilibrium. At the same time, the preheater was also set to 180°C. Then, the toluene was injected into the system at a flow rate of 0.04 ml/min for 4 hours. Once the reaction is over, the toluene flow was stopped, the furnace was set to room temperature, and the Argon flow was set to 100 ml/min. The product was collected once the furnace was cooled.

The experiment was also repeated to assess the effects of hydrogen in influencing the CNT growth. The hydrogen was introduced only during the time when toluene was injected. Two flow rates of hydrogen were used which are 50 ml/min and 100 ml/min. For such cased the flow rate of Argon was reduced to 250 ml/min and 200 ml/min, respectively to ensure that the residence time within the furnaces remains constant for all cases.

All samples including the spherical silica gel substrate were viewed under Zeiss Gemini Scanning Electron Microscope 500 (FESEM). A small amount of sample was placed on a conductive and adhesive carbon tape, which was placed on an aluminum stub. The image was viewed under high vacuum, without undergo coating. Besides FESEM, the sample was also subjected for High Resolution Transmission Electron Microscopy (HRTEM), which was also performed to view the tube's inner diameter (ID) and the outer diameter (OD) as well as wall formation. The samples were viewed under bright field mode by using LIBRA 200FE. Thermo Gravimetric Analysis (TGA) was also used to determine the amount of the CNTs produced as well as the oxidation temperature, which indicates the thermal stability of the sample. Simultaneous thermal analyzer (STA) 8000 by Perkin Elmer was used in this work. Approximately 10 ± 0.5 mg particle was placed in the pan sample holder before heating it (from 30°C to 1000 °C at 5°C/min) in oxygen atmosphere at a flow rate of 20 ml/min. Raman Spectrometry (inVia Contor Raman microscope by Renishaw) was also used to confirm the

IOP Conf. Series: Journal of Physics: Conf. Series **829** (2017) 012013 doi:10.1088/1742-6596/829/1/012013

crystallinity of the samples produced. The laser wavelength was set to 532 nm with 10s of exposure time and 10% laser power.

3. Results and Discussion

Figure 1 below show the FESEM images captured for all samples viewed at relatively low (left) and high (right) magnification. Without H_2, the CNT density seems higher (Fig 1a) and the tubes appear thicker (Fig 1b). Increasing the H_2 flow rate to 50 ml/min and 100 ml/min resulted in the formation of thinner tubes (Fig 1d and 1f). However, it should be mentioned that the CNT density seems to reduce slightly for samples produced with H_2 atmosphere (Fig 1c and 1e).

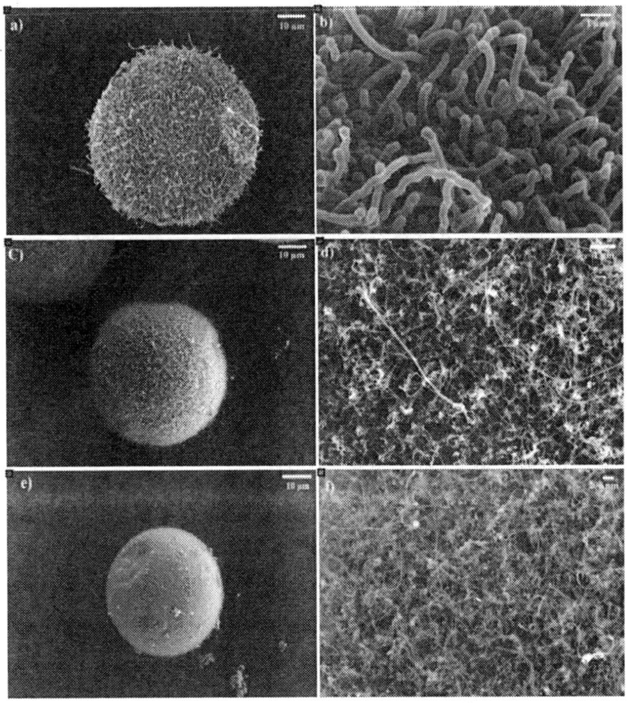

Figure 1. Typical FESEM images of CNT synthesized on spherical silica substrate under 0 ml/min H_2 (a and b), 50 ml/min H_2 (c and d), and 100 ml/min H_2 (e and f).

It can be seen immediately that the tubes synthesized without H_2 flow appear thicker (Fig 1b), compared to the those synthesized under 50 ml/min H_2 flow rate (Fig 1d) and under 100 ml/min H_2 flow rate (Fig 1f). It could also be seen that the density of the CNTs seemed to be higher when no H_2 was introduced. It is also obvious that the OD of the tubes synthesized under H_2 atmosphere is smaller, especially at 100 ml/min H_2 flow rate.

Typical images obtained from HRTEM analysis are as shown in Figure 2. It becomes apparent that the tubes synthesized in the absence of H_2 flow are very thick due to the deposition of amorphous carbon coating the wall (Fig 2a and 2b). This indicates excessive deposition of pyrolitic amorphous carbon during these 4 hours of growth time. However, it can be seen that the outer diameter (OD) of the tubes synthesized under hydrogen flow is smaller (Fig 2c and 2d). A well graphitic wall was also observed in the samples synthesized in the presence of hydrogen at both flow rates. The outer walls also appear to be free from amorphous coating, demonstrating the role of H_2 in etching the unwanted amorphous carbon deposition on the outer walls, causing a formation of thin OD of the synthesized CNTs.

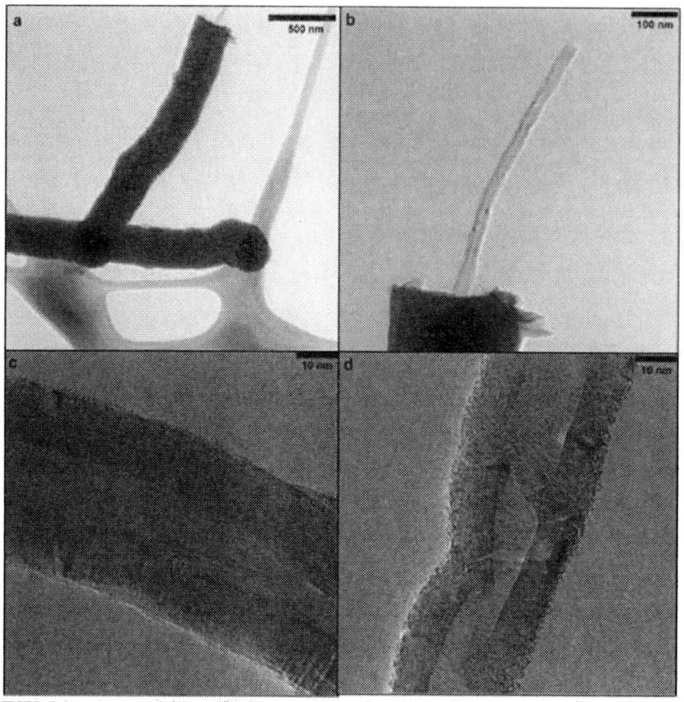

Figure 2. Typical TEM images of the CNTs produced without hydrogen flow (a and b), 50 ml/min H_2 (c), and 100 ml/min H_2 (d).

Further investigation on the inner diameter (ID) size shows that values do not change drastically regardless whether the H_2h is introduced or not (Fig 3). The mean ID values recorded are 27.46±1.80 nm (no H_2), 10.44±2.93 nm (50 ml/min H_2), and 9.29±2.35 nm (100 ml/min H_2); which indicates that H_2 flow does not exert any effects on the ID size of the tubes. FESEM images of the uncoated porous substrate revealed that there exist pores with sizes within the range of the mean ID of the tubes. In the CNT growth via CVD, the catalyst that was already placed within the heating zone first decomposed into iron and deposited over the porous substrates. Then, the toluene decomposed on the catalyst to yield carbon and hydrogen atoms. These carbon atoms underwent surface diffusion along the catalyst surface before forming a tubular structure [13]. It can be concluded in this study that the pore size restricts the catalyst size, which eventually restricts the ID of the tubes.

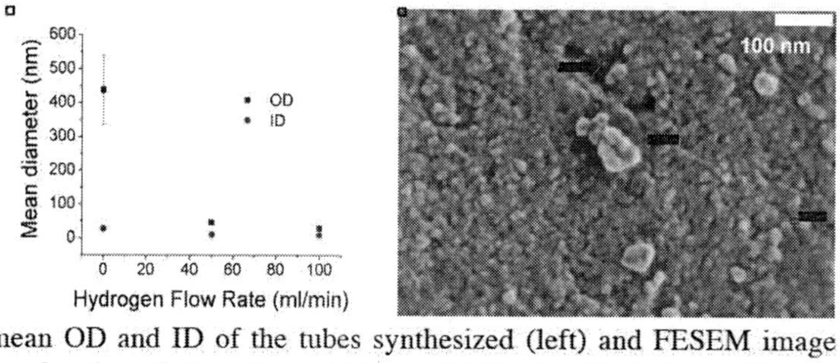

Figure 3. The mean OD and ID of the tubes synthesized (left) and FESEM image of the uncoated substrate showing the size of the pores which are 30.23 nm, 12.68 nm, and 8.22 nm from top to bottom, respectively (right).

In TGA analysis, the particles were heated at a fixed heating rate from room temperature to 1000°C in oxygen environment. The CNTs would start to decompose between 450 – 550°C. As such the weight loss shown in Figure 4 represent the amount of CNT grafted on the substrate. Presence of H_2 decreased the CNT amount, with the sample produced at 100 ml/min H_2 resulting in the lowest CNT amount. Although the CNT amount is the highest when no H_2 is present, the quality is the poorest as evidenced from FESEM and TEM analysis.

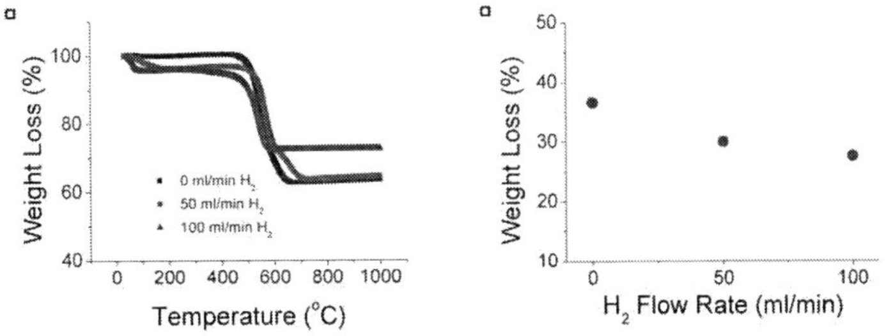

Figure 4. TG curves for samples produced (left) and the corresponding carbon content (right).

Raman analysis provides a quick, nondestructive technique in assessing the crystallinity of the CNT. The G band is the characteristic peak for the stretching vibration of carbon sp2 bonds in a hexagonal lattice, and the D band corresponds to the disordered sp2 phase plus possible contributions of the scattering of sp3-bonded carbon. It can be seen that all specimens display peaks near 1336 and 1590 cm^{-1}, assigned to the D- and G-bands, respectively (Fig 5 (left)). The D peak is very weak in the case for samples produced under 100 ml/min H_2 flow, indicating CNTs produced is of low defects. The peak intensity ratio of the G band to the D band (I_G/I_D) is also shown that indicates the CNT crystallinity (Fig 5(right)). In the presence of hydrogen flow, the ratio is more than 2, indicating high crystallinity of the CNTs produced. The value is highest in the case of 100 ml/min H_2. H_2 etched the amorphous materials, minimizing the defects on CNT walls and hence increasing the I_G/I_D ratio.

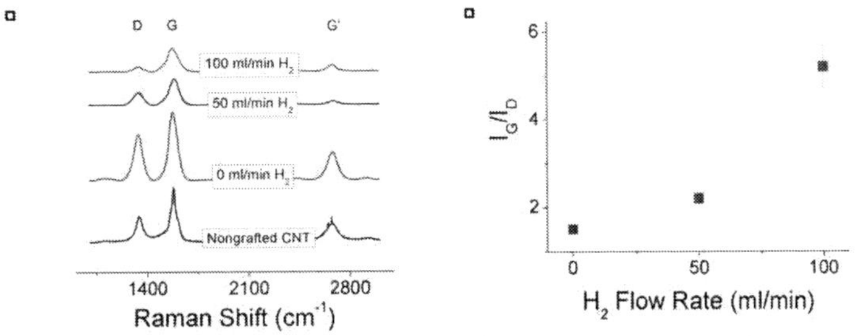

Figure 5. Raman Spectra for sample of different hydrogen concentrations where the spectra for nongrafted CNT (purchased) is shown for comparison (left) and the corresponding I_G/I_D ratio (right).

4. Conclusions

In this work, CNTs were successfully synthesized on porous spherical silica substrate via CVD method. The mean ID follows the pore size of the substrate, regardless of the H_2. The mean OD shows an obvious difference; where higher H_2 flow rate causes thinner OD formation, demonstrating role of H_2 in etching the outer wall of the tubes. Raman results confirm high crystallinity of the samples produced at high H_2 flow rate.

Acknowledgement
Financial support from both FRGS/2/2013/TK04/UPNM/03/1 and RACE/F2/TK/ UPNM/1 grants are acknowledged.

References

[1] Sandler JKW, Kirk JE, Kinloch IA, Shaffer MSP, and Windle AH, 2001 *Polymer* **44** 5893
[2] Ryan KP, Cadek M, Nicolosi V, Blond D, Ruether M, and Armstrong G, 2007 *Compos. Sci. Technol.* **67** 1640
[3] Islam MF, Rojas E, Bergey DM, Johnson AT, and Yodh AG, 2003 *Nano Lett.* **2** 269
[4] Godara A and Vaugien T, *JEC Composites*, 2009 64
[5] Othman RN, Kinloch IA, and Wilkinson AN, 2013 *Carbon* **60** 461
[6] Qian H, Bismarck A, Greenhalgh ES, Kalinka G, and Shaffer MSP 2008 *Chem Mat.* **20** 1862
[7] Qian H, Bismarck A, Greenhalgh ES, and Shaffer MSP 2010 *Carbon* **48** 277
[8] He DL, Bozlar M, Genestoux M, and Bai JB 2010 *Carbon* **48** 1159
[9] Wilkinson AN, Kinloch IA, and Othman RN 2016 *Polymer* **98** 32
[10] Huang SM 2003 *Carbon* **41** 2347
[11] Zhang H, Cao G, Wang Y, Yang Y, Shi Z, and Gu Z 2008 *J. Phys. Chem. C* **112** 12706
[12] Wasel W, Kuwana K, Reilly PTA, and Saito K 2007 *Carbon* **45** 833
[13] Oberlin A, Endo M, and Koyama T 1976 *J Cryst Growth* **32** 335

Green Synthesis of ZnO Nanoparticles by an Alginate Mediated Ion-Exchange Process and a case study for Photocatalysis of Methylene Blue Dye

Choo Cheng Keong[1], Yamini Sunitha Vivek[1], Babak Salamatinia[1,*], Bahman Amini Horri[2,*]

[1]Discipline of Chemical Engineering, School of Engineering, Monash University Malaysia, Jalan Lagoon Selatan, 46150 Bandar Sunway, Selangor, Malaysia.
[2]Department of Chemical Engineering, Faculty of Engineering, University of Surrey, Guildford GU2 7XH, United Kingdom

E-mail: babak.salamatinia@monash.edu

Abstract. In this study, zinc oxide (ZnO) was prepared via extrusion-dripping method through an ion exchange mediated process using sodium alginate. The samples were synthesized at 500 °C and 600 °C to study the effect of calcination temperature. The morphology, microstructure and optical activity of the calcined ZnO nanoparticles were analyzed by TGA, FESEM and XRD. It was found that ZnO nanoparticles synthesized at 600 °C was of higher purity with high crystallinity. To enhance the photocatalytic efficiency of zinc oxide, ZnO/NCC films were synthesized at varying ZnO loading fractions of 10 wt%, 15 wt%, 20 wt% and 25 wt% and were evaluated by photodegradation of Methylene blue dye and the highest dye percentage removal is found to be 96% which is obtained at ZnO loadings of 25 wt%. The usage of ion-exchange process has shown promising results in producing ZnO of desirable characteristics.

1. Introduction

Over the past decades, water pollution has become one of the major issues as it contains heavy ions and reactive dyes from industries. The huge discharge amount of waste water has driven researchers to solve this environmental problem. Conventional methods such as precipitation, air stripping, adsorption, reverse osmosis and ultra-filtration are non-destructive techniques as they only transfer the non-biodegradable matter into sludge [1] where only phase change occurs. This leads to pollution and the need for further treatment arises which leads to additional costs.

Zinc Oxide (ZnO) is an n-type semiconductor notable for its wide band gap (3.2 eV), large binding energy (60 meV), non-toxic nature and similar photodegradation mechanism to TiO_2 [2] which allows it to be a potential alternate photocatalyst. Many techniques have been employed to synthesis nanosized ZnO, namely sol-gel processing [3], co-precipitation [4], hydrothermal growth [5] and thermal decomposition [6]. Despite the importance of emphasizing on the geometry, size and chemical homogeneity, the overall feasibility and simplicity of the ZnO synthesis processes are also greatly considered to improve the efficiency of the fabrication process. Most of these synthesizing techniques are usually associated with energy consuming and multi-step time methods which are not environmentally friendly [7].

Extrusion dripping is a novel technique which mainly comprises of the extruding of desired solution through a narrow passageway and dripping it into a gelation bath. Mono-dispersed spherical particles are formed when the liquid drop enters the gelation bath due to effect of viscous-surface tension forces and impact-drag forces which preserves the drop shape [8]. Previous study on the preparation of spherical Ca-Alginate beads encapsulated with oil was successfully performed through extrusion dripping by Chan [9] where a stable alginate-oil emulsion was obtained. Furthermore, spherical chitosan-HNT beads were successfully produced through extrusion dripping which shows a high tendency of crosslinking copper ions [10]. Various modifications have been extensively researched in the recent years on synthesis of ZnO nanoparticles as its properties provide promising application for photocatalysis. Although ZnO finds a wider application in terms of availability, low cost and favorable gap energy, it possesses certain drawbacks as a semiconductor photocatalyst, namely the difficulty in the separation of ZnO powder after the degradation reaction, formation of aggregation at high loading and difficulty in application to continuous flow systems [11].

Cellulose, the most abundant biopolymer, is a potential alternative support for fabricating nanocomposites with semiconductors due to its thermal stability and insolubility in ordinary solvents [12]. As a renewable biomass material, cellulose is degradable, environmental friendly and cost effective. Cellulose is a linear polysaccharide with long chains which consists of β-D-glucopyranose units joined with β-1,4 glycosidic links. The repeating chains of cellulose contain methylol and hydroxyl groups as its functional group. The high amount of active sites present in the structure of cellulose and high cation-exchange capacity of its functional groups enables the coordination of metal ions on cellulose [13].

The improved thermal, mechanical stability and solvent resistibility of the hybrid in the presence of cellulose leads to a recyclable photocatalyst. However, the incorporation of support to obtained porous inorganic metal oxides is difficult due to the high reactivity of transition metal which complicates the synthesis process [14]. In this study, zinc oxide nanoparticles will be synthesized via an extrusion dripped ion-exchange mediated process and templated on nanocrystalline cellulose (NCC) as a new polymerizable support. The application of NCC as a green material allows a more eco-friendly approach to the conventional synthesizing method. The effects of calcination temperature on the characteristics of the synthesized zinc oxide were studied before subjected to application tests by analyzing its photocatalytic activity in removing Methylene Blue dye in aqueous solution.

2. Materials and Methods

2.1. *Materials*

Zinc nitrate tetrahydrate (Zn(NO3)2.4H2O) with molecular weight of 261.44 g/mol and sodium alginate were purchased from Merck and Sigma-Aldrich respectively. Furthermore, nanocrystalline cellulose with sulphur content of 0.89 wt% was obtained from University of Maine. All the chemicals used in the study were of Analytical Grade.

2.2. *Preparation of ZnO nanoparticle*

2.0 wt.% sodium alginate solution was prepared by mixing with deionised water under a digital mechanical stirrer (IKA-RW 20, IKA® Works, Malaysia). Similarly, stoichiometric amount of zinc nitrate solution was prepared by dissolving zinc nitrate tetrahydrate in deionised water. ZnO nanoparticles were then synthesized by extrusion dripping by drop wise addition of 10 mL alginate solution, controlled using syringe pump to 6.25 mL of zinc nitrate solution under magnetic stirring at 200rpm. The suspension was then stirred slowly for 15 hours to ensure the ion-exchange process and stabilize the nanoparticles formed. The nanoparticles were then separated from the solution through a sieve and washed with deionised water. Then, the nanoparticles were collected and oven-dried at 40°C for 24 hours prior to calcinations in furnace at temperature 500°C and 600 °C for 4 hours with a heating rate of 1°C/min. The morphology of the prepared ZnO powder is then analysed through FE-SEM (Hitachi Ultra-high resolution SU8010) by subjecting to a platinum coater (Quarum Q150R S).

IOP Conf. Series: Journal of Physics: Conf. Series **829** (2017) 012014 doi:10.1088/1742-6596/829/1/012014

Furthermore, the crystal structure of the powder specimens was observed by using X-ray Diffraction (Bruker D8 Discover) at room temperature by subjecting to Cu-Kα radiation.

2.3. Preparation of ZnO/Nanocrystalline(NCC) microfilm

The synthesized ZnO nanopowders were used to fabricate a polymeric NCC microfilm containing dispersed ZnO nanopowders. 2g of NCC was dispersed in 20 mL deionised water under an ultrasonic probe (QSONICA, 700 W, USA) where the process time was set as 6 minutes with 3 seconds pulse whereas the ultrasound amplitude is fixed at 70% under constant frequency. Different amount of ZnO nanopowders were then calculated and added based on various ZnO loading fractions of 10wt%, 15wt%, 20wt% and 25wt% before undergoing ultrasound for another 5 minutes to obtain the mixed slurry. The mixed product was then spread out on a glass plate to make a thin layer (10mL) before polymerizing in an oven at 90°C overnight. The ZnO/NCC microfilm is obtained after pulverizing the dried product into small pieces by a kitchen blender.

2.4. Photocatalytic study of ZnO/NCC microfilm

The degradation studies of the microfilm were conducted in a reactor fitted with a 5W lamp as an ultraviolet light source. 0.15g of ZnO/NCC microfilm with various ZnO loading fractions was added to 30 ml of 100 ppm Methylene Blue solution and left to be magnetically stirred during the photodegradation. The concentration of MB was then analysed at given intervals by monitoring the absorption peak at 665 nm using a UV-vis spectrophotometer.

3. Results and Discussion

3.1. Microstructural Analysis

The FESEM images shown in Figure 1A and 1B reveals the surface morphology of the nanoparticles and the effect of calcination temperature. It is evident that ZnO nanoparticles obtained from this study were spherical shaped and relatively uniform in size. However, nanoparticles synthesized at 500°C exhibits moderate agglomeration with lesser spherical particles. This is due to the low hydrolysis rate of precursor at low calcination temperature which leads to the formation of agglomerates due to a large difference in growth times between the non-aggregated nuclei and aggregated nuclei. The growth times between nuclei decreases as the calcination temperature is increased which leads to less agglomerates at 600 °C [15]. Furthermore, Figure 1C shows the FESEM image of the surface of the ZnO/NCC film after overnight oven drying. It was observed that the obtained film possesses a considerably smooth surface with little agglomerations.

Figure 1: FE-SEM images of ZnO nanoparticles calcinated at (a) 600°C, (b) 500°C and (c) ZnO/NCC microfilm (25wt%)

3.2. X-ray Diffractogram Analysis

Figure 2 shows the XRD peaks obtained for zinc oxide nanoparticles calcined at 500 °C and 600 °C. It was observed that the product shows Bragg reflections of (1 0 0), (0 0 2), (1 0 1), (1 0 2), (1 1 0), (1 0 3), (1 1 2) and (2 0 1) at 31.5°, 34°, 36°, 47°, 56.1°, 61.77°, 67.48° and 68.15° which indicates the formation of ZnO with wurtzite hexagonal phase [16]. As seen from Figure 2, it is evident that the peaks become more intense and sharper at higher calcination temperatures. The presence of high and narrow shaped peaks highlight that ZnO nanoparticles possess high crystallinity and low surface defects [17]. Furthermore, no diffraction peaks from other impurities were observed in the XRD diffractograms which indicates that the ZnO product obtained after the calcination were of ultrapure phase.

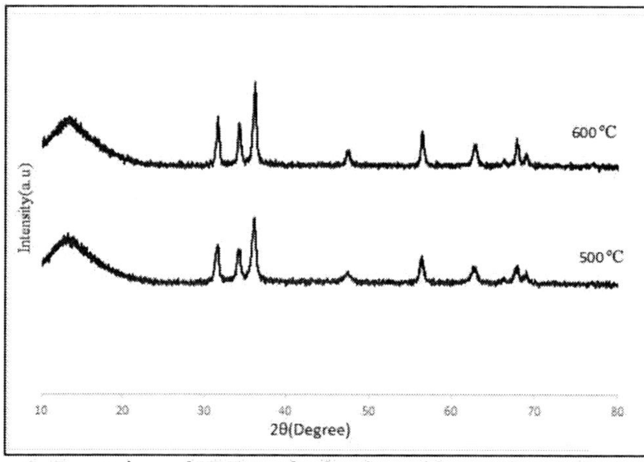

Figure 2: Comparison of XRD results for zinc oxide samples synthesized at (a) 600°C and (b) 500°C

3.3. Thermalgravimetric Analysis (TGA)

Thermogravimetric analysis (TGA) was conducted on the bead samples prior to calcination to determine the suitable starting temperature for the calcination process. Figure 3 shows the relationship between percentage weight change and temperature for the sample before calcination. By analysing the graph, it is evident that there are two major degradation steps involved. The first drop in weight loss takes place at 170 °C, which could be the loosely bond water or moisture entrapped within the sample. The second drop demonstrates a major weight loss at 390°C, and this could be attributed to the near complete decomposition of sodium alginate, which is the precursor used in this study. At 445 °C, the change in weight loss has become insignificant, and most of the sodium alginate present has been converted to carbonaceous products, leaving behind pure metal oxide as residue [18]. It is evident that the organic compound has completely decomposed after 450°C, therefore 500 °C was selected as the lower limit of calcination temperature of the extrusion dripped beads.

IOP Conf. Series: Journal of Physics: Conf. Series **829** (2017) 012014 doi:10.1088/1742-6596/829/1/012014

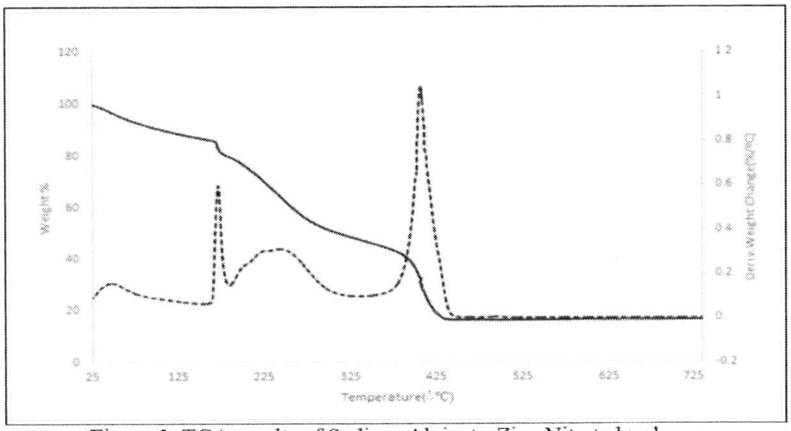

Figure 3: TGA results of Sodium Alginate-Zinc Nitrate beads

3.4. Photocatalytic Application

Figure 4 shows the degradation behaviour of methylene blue in ZnO/NCC microfilm loaded with 15wt%, 20wt% and 25wt% ZnO loadings. The percentage removal of methylene blue dye was calculated using the following equation:

$$\text{Percentage Removal} = \frac{C_o - C_T}{C_o} \times 100 \qquad (1)$$

where C_o is the initial concentration of dye (mg/L) and C_T is the concentration of dye (mg/L) at different contact times. It is very evident from the trend of the graph that the percentage removal of dye is increased with increasing weight loadings of ZnO where the highest percentage dye removal were found to be 96% using 25wt% ZnO loading. As seen in Figure 1C, it is clear that the ZnO particles are densely residing on the outer surfaces of the film where it plays a major role in the photodegradation process as they accept photons and generate electron- holes for the oxidative decomposition of methylene blue. The enhancement in photodegradation activity which is observed at high ZnO loading could be attributed to the increased in total surface area and total number of active sites on the catalyst surface. This leads to the increase in the number of hydroxyl and superoxide radicals in the catalyst material which facilitates the overall degradation of methylene blue dye [19].

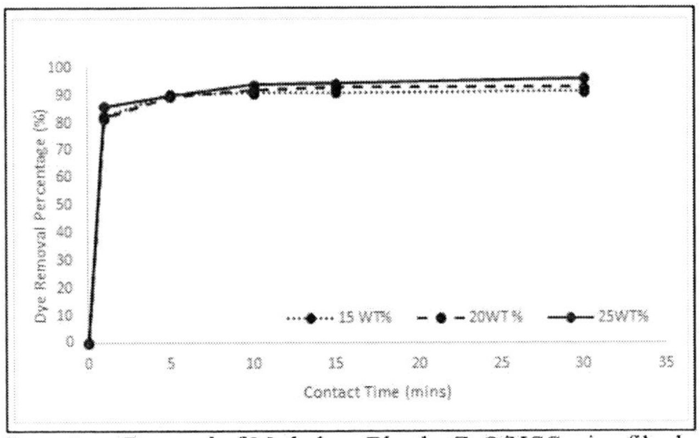

Figure 4: Percentage Removal of Methylene Blue by ZnO/NCC microfilm loaded with (a) 15wt%, (b) 20wt% (c) 25wt% loading fraction

4. Conclusion

Successful synthesis of the ZnO nanoparticles was achieved using an alginate mediated ion-exchange process for the preparation of a ZnO/NCC microfilm where the photodegradation studies of methylene blue dye is conducted as a case study. The nanoparticles formed at the calcination temperature of 600°C were found to provide the most desirable characteristics as shown from the results obtained from FE-SEM and XRD. Furthermore, the microfilm with 25wt% ZnO loading showed the highest methylene dye removal where the percentage removal is found to be 96%. The application of green materials provides a more eco-friendly approach to the current technologies in wastewater treatment.

Acknowledgement
The authors would like to acknowledge Ministry of Education Malaysia (MOE) for providing research funding under the FRGS scheme number FRGS/2/2014 TK05/MUSM/02/1. Furthermore, the authors would like to thank Monash University Malaysia for all the analytical support provided. Besides that, Ms Indi Chandra Harrisa is also highly appreciated for her assistance throughout the project.

References
[1] P. Muthirulan, M. Meenakshisundararam, N. Kannan, Beneficial role of ZnO photocatalyst supported with porous activated carbon for the mineralization of alizarin cyanin green dye in aqueous solution, Journal of advanced research, 4 (2013) 479-484.
[2] J. Xie, Y. Li, W. Zhao, L. Bian, Y. Wei, Simple fabrication and photocatalytic activity of ZnO particles with different morphologies, Powder Technology, 207 (2011) 140-144.
[3] Y. Hou, A.H. Jayatissa, Low resistive gallium doped nanocrystalline zinc oxide for gas sensor application via sol–gel process, Sensors and Actuators B: Chemical, 204 (2014) 310-318.
[4] X. Zhao, M. Li, X. Lou, Enhanced photocatalytic activity of zinc oxide synthesized by calcination of zinc sulfide precursor, Materials Science in Semiconductor Processing, 16 (2013) 489-494.
[5] A. Syed, M. Kalloudis, V. Koutsos, E. Mastropaolo, Controlled hydrothermal growth of vertically-aligned zinc oxide nanowires using silicon and polyimide substrates, Microelectronic Engineering, 145 (2015) 86-90.
[6] I. Kontopoulou, A. Angelopoulou, N. Bouropoulos, ZnO spherical porous nanostructures obtained by thermal decomposition of zinc palmitate, Materials Letters, 165 (2016) 87-90.
[7] A.B. Sifontes, G. Gonzalez, J.L. Ochoa, L.M. Tovar, T. Zoltan, E. Cañizales, Chitosan as template for the synthesis of ceria nanoparticles, Materials Research Bulletin, 46 (2011) 1794-1799.
[8] E.S. Chan, B.B. Lee, P. Ravindra, D. Poncelet, Prediction models for shape and size of ca-alginate macrobeads produced through extrusion-dripping method, Journal of colloid and interface science, 338 (2009) 63-72.
[9] E.-S. Chan, Preparation of Ca-alginate beads containing high oil content: Influence of process variables on encapsulation efficiency and bead properties, Carbohydrate polymers, 84 (2011) 1267-1275.
[10] C.K. Choo, X.Y. Kong, T.L. Goh, G.C. Ngoh, B.A. Horri, B. Salamatinia, Chitosan/halloysite beads fabricated by ultrasonic-assisted extrusion-dripping and a case study application for copper ion removal, Carbohydrate polymers, 138 (2016) 16-26.
[11] P. Muthirulan, C. Nirmala Devi, M. Meenakshi Sundaram, Synchronous role of coupled adsorption and photocatalytic degradation on CAC–TiO2 composite generating excellent mineralization of alizarin cyanine green dye in aqueous solution, Arabian Journal of Chemistry, (2013).
[12] S. Ye, D. Zhang, H. Liu, J. Zhou, ZnO nanocrystallites/cellulose hybrid nanofibers fabricated by electrospinning and solvothermal techniques and their photocatalytic activity, Journal of Applied Polymer Science, 121 (2011) 1757-1764.
[13] S. Hokkanen, A. Bhatnagar, M. Sillanpaa, A review on modification methods to cellulose-based adsorbents to improve adsorption capacity, Water research, 91 (2016) 156-173.
[14] N. Dharmaraj, P. Prabu, S. Nagarajan, C.H. Kim, J.H. Park, H.Y. Kim, Synthesis of nickel oxide nanoparticles using nickel acetate and poly(vinyl acetate) precursor, Materials Science and Engineering: B, 128 (2006) 111-114.
[15] A. Ishida, S. Hori, T. Tani, T. Ikeda-Fukazawa, M. Aizawa, Hydrothermal synthesis of single-crystal α-tristrontium phosphate particles, Journal of the European Ceramic Society, 37 (2017) 351-357.
[16] F. Ghasemi, R. Jalal, Antimicrobial action of zinc oxide nanoparticles in combination with ciprofloxacin and ceftazidime against multidrug-resistant Acinetobacter baumannii, Journal of global antimicrobial resistance, 6 (2016) 118-122.
[17] O. Lupan, G.A. Emelchenko, V.V. Ursaki, G. Chai, A.N. Redkin, A.N. Gruzintsev, I.M. Tiginyanu, L. Chow, L.K. Ono, B. Roldan Cuenya, H. Heinrich, E.E. Yakimov, Synthesis and characterization of ZnO nanowires for nanosensor applications, Materials Research Bulletin, 45 (2010) 1026-1032.
[18] K. Ada, M. Gökgöz, M. Önal, Y. Sarıkaya, Preparation and characterization of a ZnO powder with the hexagonal plate particles, Powder Technology, 181 (2008) 285-291.
[19] K.M. Lee, C.W. Lai, K.S. Ngai, J.C. Juan, Recent developments of zinc oxide based photocatalyst in water treatment technology: A review, Water research, 88 (2016) 428-448.

Electrospun Nanocomposite Materials, A Novel Synergy of Polyurethane and Bovine Derived Hydroxyapatite

Y Bozkurt[1*], A Sahin[2,3], A Sunulu[2,3], M O Aydogdu[2,4], E Altun[2,4], F N Oktar[2,5], N Ekren[2,6] and O Gunduz[1,2]

[1] Department of Metallurgy and Materials Engineering, Faculty of Technology, Marmara University, Goztepe Campus, 34722, Istanbul, Turkey
[2] Advanced Nanomaterials Research Laboratory, Department of Metallurgical and Materials Engineering, Marmara University, Goztepe Campus 34722 Istanbul, Turkey
[3] Bioengineering Department, Bachelor's Degree, Faculty of Engineering, Marmara University, Goztepe Campus, 34722, Istanbul, Turkey
[4] Department of Metallurgy and Materials Engineering, Master of Science, Institute of Pure and Applied Sciences, Marmara University, Goztepe Campus 34722 Istanbul, Turkey
[5] Bioengineering Department, Faculty of Engineering, Marmara University, Goztepe Campus, 34722, Istanbul, Turkey
[6] Department of Electric-Electronic Engineering, Faculty of Technology, Marmara University, Goztepe Campus 34722, Istanbul, Turkey

E-mail: ybozkurt@marmara.edu.tr*, ahmet_1780@hotmail.com, aknsunulu@gmail.com, aydogdumonur@gmail.com, esra.altun@ymail.com, foktar@marmara.edu.tr, nazmiekren@marmara.edu.tr, oguzhan@marmara.edu.tr.

Abstract. Polyurethane (PU) is a synthetic polymer that is used for construction of scaffold in tissue engineering applications in order to obtain desirable mechanical, physical and chemical properties like elasticity and durability. Bovine derived hydroxyapatite (BHAp) is a ceramic based natural polymer that is used as the most preferred implant material in orthopedics and dentistry due to their chemically and biologically similarity to the mineral phase found in the human bone structure. PU and bovine derived hydroxyapatite (BHAp) solutions with different concentrations were prepared with dissolving polyurethane and BHAp in Dimethylformamide (DMF) and Tetrahydrofuran (THF) solutions. Blended PU-BHAp solutions in different concentrations were used for electrospinning technique to create nanofiber scaffolds and new biocomposite material together. SEM, FTIR and physical analysis such as viscosity, electrical conductivity, density measurement and tensile strength measurement tests were carried out after production process.

1. Introduction

Tissue engineering deals with the creation of tissue scaffolds in the laboratory conditions for implanting organs and tissues [1]. Tissue engineering, physical and chemical aspects of bone extracellular matrix

(ECM) that may mimic (biomimicry) biomaterial (tissue scaffold), aims to repair or regenerate the damaged tissue or not functional ones in any part of human body, using the bone formation components of stimulating molecules and the cells with the potential to create under appropriate conditions [2].

Tissue engineering methods can be generally described as cell adhesion and preserving the functionality of a tissue support that supports a rich source of cells selected according to the target tissue and include growth factors that control the behavior of these cells [3].

The materials used in tissue scaffolds should be biocompatible and not to lead undesirable tissue response also it should have a surface chemistry increasing the cell adhesion and function. Tissue scaffolds fabricated from biodegradable material which has degradation ability in the body [4]. In addition to these features, biomaterials should be processable for the production of damage-specific structure; it can be recycled to the physical integrity of the bone ıf tissue scaffold is expected to have the desired mechanical strength. Providing these features depends on selected tissue scaffold production method and used biomaterials. Tissue scaffold should be possessed specific feature that can be listed as follows: Biocompatibility, biodegradability, mechanical properties, scaffold architecture and manufacturing technology [5].

Hydroxyapatite (HA), $Ca10(PO_4)6(OH)_2$ is a ceramic based polymer that is used as the most preferred implant material in orthopedics and dentistry due to the chemically and biologically similarity to the mineral phase found in the human bone structure. BHAp which is possessed of rhombic hexagonal lattice structure, plays an active role ensuring the bone mineral balance and maintaining the mechanical strength but it is brittle failure for that reason bovine derived BHAp is combined with PU in this way mechanical properties of resulting composite can be improved while establishing interfacial bond strength between the ceramic phase and the polymer matrix [6,7].

Polyurethanes are synthetic polymers, used for construction of scaffold due to their flexibility and versatile chemistry [8]. Biocompatible and biodegradable polyurethanes are used as suitable scaffold for tissue engineering in order to obtain desirable mechanical, physical and chemical properties like elasticity and durability [9]. Polyurethane is one of the most widely used polymers that are used in biomedical applications that contacts with blood. As a result of this, it can be easily synthesized in different forms and surfaces as compared to other polymers would not coagulated. Current applications of polyurethanes are catheters, blood bags and artificial heart systems etc.

Electrospinning technique is an efficient and relatively easy polymer fabrication technique that is allowed to produce wide range of nanofibers from various polymers by using voltage to create an electric field [10].

Previous research has been investigated for electrospun polyurethane/hydroxyapatite scaffolds [11], development and characterization of Polyurethane-Hydroxyapatite bio nanocomposites [12], Polyurethane/fluorhydroxyapatite nanocomposite scaffolds [13], investigation of PU and HA [14].

Aim of this study is to improve the bioactive properties of BHAp and PU nanofibers compared to only-polymer scaffolds while establishing interfacial bond between each other so that preferable mechanical properties of nanofibers were acquired with using that biomaterials together. We also investigated the mechanical properties of BHAp particles and PU nanofibers with different tests respectively; electrical conductivity, density, viscosity, tensile measurement, scanning electron microscopy (SEM) and Fourier-Transform Infrared Spectroscopy (FTIR).

2. Experimental process

2.1. Materials and chemicals

Solvents named Tetrahydrofuran (THF), Dimethylformamide (DMF) were supplied from Sigma - Aldrich (St. Louis, MO, USA) and PU were purchased from TERMOSAN (Istanbul, Turkey) used without distillation or pre-treatment. BHAp were prepared in the nano sized diameters, derived from bovine bone.

2.2. Methods

2.2.1. Solution preparing and electrospinning process. PU was dissolved in different concentrations with THF/DMF (70:30 wt. %) solvents. Magnetic stirrer (WiseStir®, MSH-20A, Deutschland) were used to speed up the dissolving process. BHAp were added to the selected solutions and magnetic stirring process were repeated for 8 hours at 50 °C. Pure PU and PU-BHAp blend solutions were prepared in twelve different concentrations with the purpose of determining the best composite components to obtain novel and enhanced features. Contents of the samples were shown together with electrospinning parameters in Table 1.

Table 1. Nomenclature for the samples used with their contents and parameters.

Abbreviation of the samples prepared	Content		Parameters	
	PU (wt. %)	BHAp (wt. %)	Voltage (kV)	Flow rate (ml/hr)
s1	3	0	27	1
s2	5	3	21.5	1.1
s3	7	3	32.4	1.3
s4	14	3	21	1.5
s5	3	3	26.4	0.3
s6	5	3	27.4	0.7
s7	7	3	23.6	0.6
s8	14	3	18.6	0.4
s9	3	5	23	0.6
s10	5	5	18	0.3
s11	7	5	26.2	0.7
s12	14	5	24	0.9

2.3. Characterization

Characterization properties of Pure PU and PU-BHAp blend solutions and nanofibers were described and tests were done in order to get appropriate scaffolds.

2.3.1. Viscosity electrical conductivity and density of the solutions. Viscosity measurement of prepared Pure PU and PU – BHAp blend solutions were processed with Rheometer Device (Manufacturer- M/s Malvern Instruments Ltd., U.K., Model: Bohlin Gemini Rheometer) at room temperature, range between 1-200 Pa.s. Electrical conductivity tests were carried out with (WTW, Cond 3110 SET1, Germany) related device at room temperature. Density measurements were carried out with pycnometer at room temperature and all results were shown in Table 2.

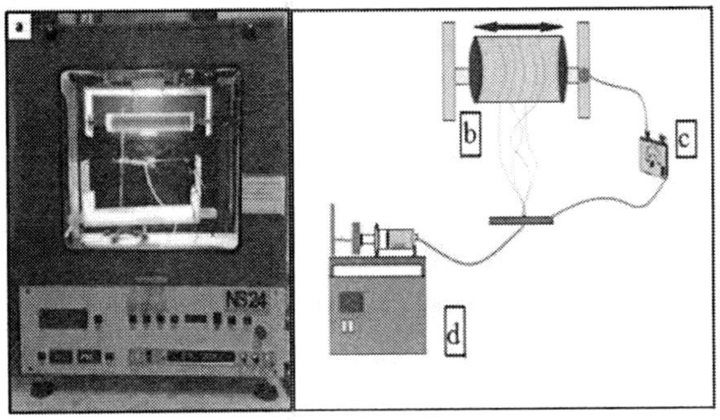

Figure 1. Camera image of the electrospinning device (a) and its schematic design with rotating collector (b), regulated power supply (c), digital syringe pump (d).

2.3.2. Mechanical properties of nanofibers. Mechanical characteristics of the nanofibers obtained were tested by the tensile strength measurement device (INSTRON 4411, Massachusetts, USA) at room temperature (23 ° C) adjusted speed were set to 10 mm/min. Thicknesses of the fibers produced were analyzed with digital micrometer (795.1 MEXFL-25, Starrett, USA).

Table 2. Physical test results of the samples such as density, viscosity, electrical conductivity and tensile strength measurement tests.

Concentration of the solutions	Tensile Strength Measurement (MPa)	Electrical Conductivity (μS/m)	Viscosity (Pa.s)	Density (kg/m^3)
3 wt. % PU (s1)	0.93	1.4	6.403	0.926
5 wt. % PU (s2)	1.11	0.9	4.632	0.927
7 wt. % PU (s3)	1.15	0.8	3.059	0.929
14 wt. % PU (s4)	1.64	0.8	0.520	0.950
3 wt. % PU, 3 wt. % BHAp (s5)	0.22	9.4	4.240	0.935
3 wt. % PU, 5 wt. % BHAp (s6)	0.19	10.2	5.318	0.936
5 wt. % PU, 3 wt. % BHAp (s7)	0.69	8.1	7.241	0.939
5 wt. % PU, 5 wt. % BHAp (s8)	0.57	9.1	6.148	0.941
7 wt. % PU, 3 wt. % BHAp (s9)	0.47	9.7	3.067	0.958
7 wt. % PU, 5 wt. % BHAp (s10)	0.58	13	2.616	0.960
14 wt. % PU, 3 wt. % BHAp (s11)	0.11	8.5	0.423	0.971
14 wt. % PU, 5 wt. % BHAp (s12)	0.09	10.5	0.312	0.975

2.3.3. Scanning electron microscopy. Morphological and topographical details of prepared Pure PU and PU-BHAp electrospun fibers were analyzed with scanning electron microscopy (Zeiss Evo LS10. Germany) and tests were carried out at Yildiz Technical University. Prior to imaging, samples were coated with gold powder at 60 seconds. Coated samples were analyzed at 5 kV, a spot size of 7 and a magnification of 10000 times.

2.3.4. Fourier transform infrared spectroscopy. Content of the Pu-BHAp nanocomposite fibers were analyzed with Fourier-Transform Infrared Spectroscopy (GC-IR Interface for Nicolet™ FT-IR Spectrometers, USA). Spectroscopy was performed to determine the chemical components of the nanofibers produced between the ranges from 525 to 4000 cm^{-1}.

3. Results and Discussions

Table 1 shows the contents and concentrations of the materials used with two major parameters of the electrospinning process which are flow rate and voltage used to produce nanofibers. In Figure 1, (a) camera image of the electrospinning device can be seen in schematic design of the electrospinning device and its components such as (b) rotating collector, which increases the surface area in order to collect polymers solution on it and causes liquid to evaporate, (c) regulated power supply, which provides power to the systems (d) digital syringe pump to adjust flow rate of polymer solutions into electrical field.

On the SEM images nanofiber formations were observed and shown in Figure 2 as (a) low magnification and (b) high magnification images of the 's5'.

At %14 PU concentration nanofiber scaffolds were observed with the bead like structures which is generated by the presence of BHAp. Due to results of blending process, fiber structures and bead-like structures were seen together as one fine scaffold. Diameters of the nanofibers were observed from 126.7 nm to 175.1 nm for 's8' and from 131.2 to 154.2 nm for 's12'.

As seen on the (c) low magnification and (d) high magnification images of 's8' and (e) low magnification and (f) high magnification images of 's12' respectively at Figure 2.

When the concentration of the BHAp increased, the presence and sizes of the bead like structures and inner diameters of the fiber scaffolds were also slightly increased. Viscosity of our solutions between 7.2 and 0.3, when the concentration of PU and BHAp were increased, viscosity of solutions was decreased as shown in Table 2. Densities of solutions are between 0.92 and 0.97 kg\m^3. Electrical measurement of our solutions is between 0.8 and 13 µs/m, when the concentration of PU was increased, dielectric constants of solutions were decreased as shown in Table 2.

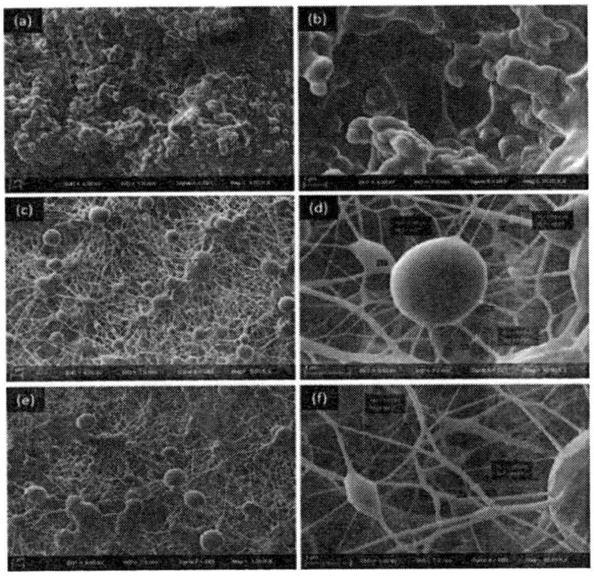

Figure 2. SEM images of the s5, s8, s12 respectively, (a,c,e) low and (b,d,f) high magnification.

FTIR spectra patterns of different concentration of PU and PU-BHAp composites nanofibers were shown in Figure 3. The peak at 3325 cm^{-1} represents N-H bonds which is characteristic for PU. The peak at 3550 cm^{-1} is related to OH bending which is characteristic for HA. The peak at 3000 cm^{-1} describes CH_2 (Methylene) which is characteristic for HA. The peak at 1510-1700 cm^{-1} regions were assign -C=O groups which is characteristic for PU. The peak at 3000 cm^{-1} represents CH_2 (Methylene) which is characteristic for HA. The peak at 1510-1700 cm^{-1} regions were assign -C=O groups which is characteristic for PU. The peak at 1500 cm^{-1} region was assigned CO groups. The peaks between 1000-1200 cm^{-1} region denote the $(PO_4)^{-3}$ group which is characteristic for HA. The bands in the 570 and 1000 cm^{-1} specify the P-O bonds of PU and 630 cm^{-1} indicates the O-H, 600 and 560 indicates the P-O bonds of HA [16-19].

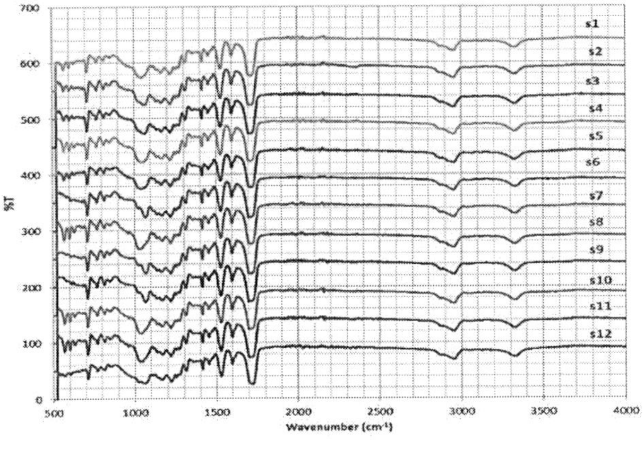

Figure 3. FTIR spectra of the obtained nanofibers.

4. Conclusions

Using the BHAp with PU was very important for ensuring biocompatibility, biodegradability, and mechanical strength of the fibrous materials. PU scaffolds were fabricated from DMF and THF solutions with THF/DMF (70:30 wt. %), which has been found to be most suitable solvent combination. Results are found to be promising for bone tissue engineering applications due to the characteristics of the PU-BHAp composite nanofiber materials.

Acknowledgement
This project is granted by Marmara University, BAPKO project numbers: FEN-D- 100816-0403.

References

[1] Atala A 2011 Tissue engineering of human bladder *Ox. J. Med. Health Brit. Med. Bull.* **97** 1:81-104
[2] Perez R et al 2015 Therapeutically relevant aspects in bone repair and regeneration *Mat. Today* **18(10)** 573-589.
[3] Chan B P and Leong K W 2008 Scaffolding in tissue engineering: general approaches and tissue-specific considerations *Eur Spine J* **17** 467-479.
[4] Dhandayuthapani B et al 2011 Polymeric scaffolds in tissue engineering application: A Review *Int. J Poly. Sci.* **2011** 1-19.
[5] O'brien, F J 2011 Biomaterials and Scaffolds for Tissue Engineering *Mat. Today* **14,** 88-95.
[6] Polo-Corrales L et al 2014 Scaffold Design for Bone Regeneration *J. Nanosci. Nanotechn.* **14(1)** 15-56.
[7] Martinez Valencia A et al 2011 Production of polyurethane/nanohydroxyapatite hybrid materials and microstructural characterisation. *Int. J. Phys. Sci.* **6** 2731-2743.
[8] Guelcher S A 2008 Biodegradable polyurethanes: synthesis and applications in regenerative medicine. *Tissue Eng. Part B: Rev.* **14** 3-17.
[9] Zdrahala R J and Zdrahala I J 1991 Biomaterial application of polyurethanes: review of past promises, present realities and a vibrant future *J. Biomater.Appl* **14** 67-90.
[10] Demir M M et al 2002 Electrospinning of polyurethane fibres. *Polymer* **24** 3303-3309.
[11] Tetteh G et al 2014 Electrospun polyurethane/hydroxyapatite scaffolds for bone tissue engineering: The role of solvent and hydroxyapatite particles, Elseiver, *J. Mech. Beh. Biomed. Mat.* **39** 95-110.
[12] Díez Pascual A et al 2014 Development and characterization of novel poly(ether ether ketone)/ZnO bionanocomposites. *J. Mat. Chem. B* **2(20)** 3065.
[13] Asefnejad A et al 2011 Polyurethane/fluor-hydroxyapatite nanocomposite scaffolds for bone tissue engineering. Part I: morphological, physical, and mechanical characterization, *Int. J. Nanomedicine* **6** 93-100.
[14] Gorna K and Gogolewski S 2003 Biodegradable porous polyurethane scaffolds for tissue repair and regeneration. *J. Biomed. Mater. Res.* **79A** 128-138.
[15] Tsui Y K and Gogolewski 2009 Microporous biodegradable polyurethane membranes for tissue engineering. *J. Mater. Sci. Mater. Med.* **20(8)** 1729-1741
[16] Bil M et al 2007 Bioactivity of polyurethane-based scaffolds coated with Bioglass *Biomed. Mater.* **2 (2)** 93-101.
[17] McCarthy S et al 1997 In-vivo degradation of polyurethanes: transmission-FTIR microscopic characterization of polyurethanes sectioned by cryomicrotomy. *Biomat.* **18** 1387-1409.
[18] Gorna K et al 2002. Synthesis and characterization of biodegradable poly (ε-caprolactone urethane) I. Effect of the polyol molecular weight, catalyst, and chain extender on the molecular and physical characteristics. *J. Polym. Sci. A Poly. Chem.* **40** 156-170.
[19] Bakan F et al. A novel low temperature Sol–gel synthesis process for thermally stable nano crystalline hydroxyapatite. Powder Technol 233:295–302.

IOP Conf. Series: Journal of Physics: Conf. Series **829** (2017) 012016 doi:10.1088/1742-6596/829/1/012016

Influnce of exposure with Xe radiation on heterojunction solar cell a-SiC/c-Si studied by impedance spectroscopy

M Perný[1], V Šály[1], J Packa[1], M Mikolášek[1], M Váry[1], J Huran[2], L Hrubčín[2,3], V A Skuratov[3] and J Arbet[2]

[1]Faculty of Electrical Engineering and Information Technology, Slovak University of Technology, Ilkovičova 3, 81219 Bratislava, Slovakia, milan.perny@stuba.sk

[2] Institute of Electrical Engineering, Slovak Academy of Sciences, Dúbravská cesta 9, 84104 Bratislava, Slovakia

[3]Joint Institute for Nuclear Research, Joliot-Curie 6, 141980 Dubna, Moscow Region, Russian Federation

Abstract. The photovoltaic efficiency of heterostructures a-SiC/c-Si may be the same or even better in comparison with conventional silicon structures when suitable adjustment of technological parameters is realized. The main advantage of heterojunction formed amorphous SiC thin film and crystalline silicon compared to standard crystalline solar cell lies in high build-in voltage and thus a high open-circuit voltage. Solar cells can be exposed to various influences of hard environment. A deterioration of properties of heterostructures (a-SiC/c-Si) due to irradiation is examined in our paper using impedance spectroscopy method. Xe ions induced damage is reflected in changes of proposed AC equivalent circuit elements. AC equivalent circuit was proposed and verified using numerical simulations. Impedance spectra were also measured at different DC bias voltages due to a more detailed understanding correlation between Xe ions induced damage and transport phenomenon in the heterostructure.

1. Introduction

Hydrogenated amorphous silicon carbide (a-SiC:H) is a wide band-gap semiconductor, alternatively modified with other chemical elements, which forms heterojunction (HJ) with crystalline silicon. It could be relevant substitute for amorphous silicon (a-Si:H) and similarly its bandgap can be tuned over the wide range. Main advantage of amorphous SiC alloys lies in possibility to be deposited in low-temperature process chemical vapor deposition (CVD) [1]. The possibility of low-temperature fabrication processes is very interesting also for photovoltaics where structures with amorphous carbide thin films have important applications. Heterojunction of bulk silicon and amorphous a-SiC thin film represents an alternative to similar heterojunction a-Si/c-Si structure. Resistance against damage in hard environmental conditions is important advantage except for optical, thermal, electrical and mechanical properties of wide gap semiconductor a-SiC. Various forms of SiC are often exploited even in devices, eg sensors, exposed to hard radiation [2] due to high bandgap and high atomic displacement energies. Application of a-SiC in photovoltaic structure lead recently to the efficiency 25.6 % achieved on heterojunction structure, which makes this technology the most efficient among the silicon based solar cells [3]. The main advantage of heterojunction formed amorphous SiC thin film and crystalline silicon compared to standard crystalline solar cell lies in high build-in voltage and thus a high open-circuit voltage [4]. Taking into account external influences PV device has to be evaluated in more complex way and all aspects of its performance shall be considered [5].

Chemical bonds in SiC are primarily covalent but small percentage of them has ionic character. The reaction of particle, like neutrons, protons or ions, flux with solid matter is through elastic collision with atomic nuclei, often resulting in the damage of crystalline structure. It is important to emphasis that some structural defects in SiC are charged, which is related to the mentioned fact that SiC is a wide band gap semiconductor [6].

We have studied heterojunction solar cell device composed of thin layer a-SiC deposited on silicon substrate. ITO layer was used for antireflection and passivation purpose which also offers adequate transmittance and high conductance.

In order to estimate the ability to withstand hard environmental condition we have studied the influence of Xe ions bombardment on prepared samples a-SiC/c-Si with ITO at the front surface. The AC electrical response and complex impedance, which showed to be convenient tool, have been investigated.

2. Experimental

The heterojunction structures were fabricated on p-type Si (100) 1 x 1 cm wafers. Phosphorus doped silicon carbide film as emitter was deposited by PECVD technology using silane (SiH_4 -10 sccm), methane (CH_4 -3 sccm), hydrogen (H_2 -100 sccm) and PH_3 (2.0 vol.% in H_2, 15-30 sccm), gas as precursors. ITO (indium oxide/tin oxide 90/10 wt. %) film was RF magnetron sputtered on top of the different P doped a-SiC:H(n) films. The thickness of the deposited ITO films was kept at 100 nm. Photolitography and lift off technique was used for Al grid preparation as top contact of solar cell structure and bottom Al ohmic contact was prepared by Al evaporation. Irradiation of structures with Xe ions to total fluency $5x10^{11}$ cm^{-2} was performed at room temperature in the IC100 accelerator at JINR Dubna, Russia. Impedance spectra were measured by the Agilent LCR Meter 4284A. The amplitude of AC signal voltage was kept at 25 mV which maintains the linearity of the response. The real and imaginary components of the impedance (Z′ and Z′′) were measured at 100 Hz to 1 MHz frequency range. DC bias was varied in the interval from - 0.8 V to + 0.7 V.

3. Results and discussion

Fig. 1 shows the obtained impedance data of the HJ structure under reverse and forward bias before and after Xe irradiation. The obtained plot has semicircular shape and are composed from superposition of two semicircles (two relaxation processes – better recognizable after irradiation). The diameter of the circles depends on the applied DC bias. Shunt resistance (diameter of semicircle) of HJ structure was significantly reduced under reverse bias conditions when Xe ion irradiation was brought into effect, as one can see in Fig. 1a.

The behavior under forward bias condition is different (Fig. 1b).

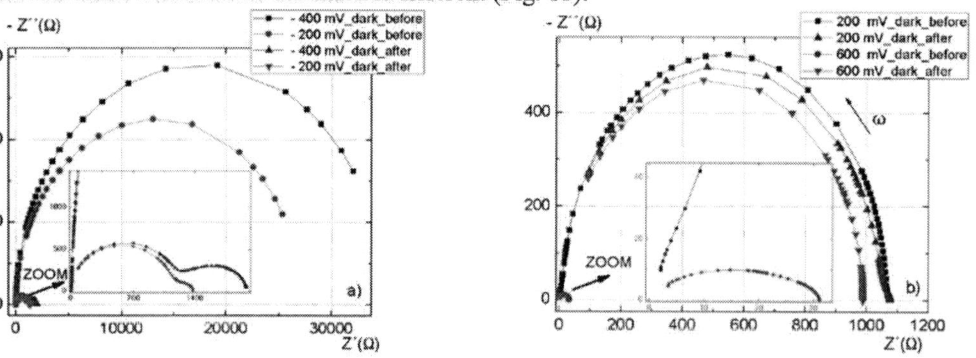

Figure 1. Obtained impedance spectra for forward and reverse conditions obtained in the dark at the room temperature.

Irradiation by Xe ions results in increasing of conductivity. This result correlates with DC measurement. The influence of Xe bombardment is demonstrated on current-voltage characteristic shown in Fig. 2.

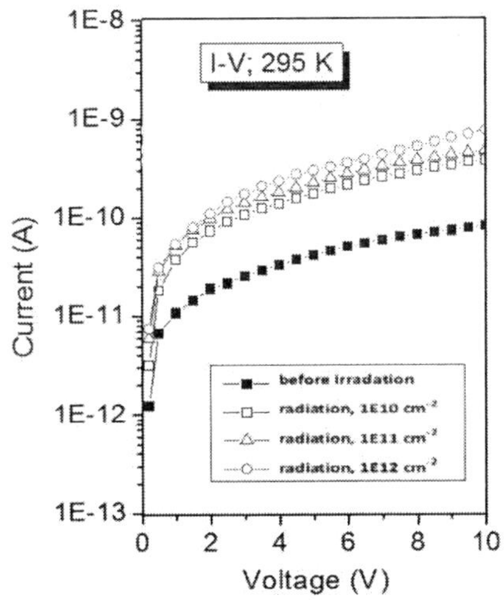

Figure 2. Example of I-V characteristic under forward bias of structure before and after Xe ions irradiation.

The increased fluency of Xe ions results in increased structure conductivity. The parallel resistance of HJ structure is combination of the shunt and dynamic resistance. Dynamic resistance of junction under forward DC bias reaches incomparable smaller value of resistance than the shunt resistance. For this reason, in forward bias conditions only dynamic resistance is contributing to the total cell resistance. Dynamic resistance decreases with increased forward bias but remains independent of frequency within certain interval of frequencies and start decrease toward series resistance value at higher frequencies [7].

Modeling of obtained impedance data by AC equivalent circuit was necessary for the understanding of processes before and after Xe irradiation inside heterostructure. Xe ions induced damage reflected in changes of AC equivalent circuit elements and equivalent circuit had to be modified. Alternative AC equivalent circuits were proposed and verified using numerical simulations. Impedance spectra were also measured at different DC bias voltages due to a more detailed understanding the correlation between ions induced damage and transport phenomenon in the heterostructure. As high energy ions penetrate into a solid, those ions can deposit kinetic energy in two processes: electronic excitation and ionization, and nuclear collision. Kinetic energy transfer between penetrating ions and target atoms takes place through collective processes of nuclear collisions in a displacement [8]. Moreover generated defects can be charged.

AC equivalent circuit, before and after Xe irradiation which best fits the obtained impedance data is shown in Fig. 3. Series resistance R_1 is related to resistance of contacts and electrode system (grid on top and whole area bottom) in both cases a) and b). Parallel combination of resistance R_2 (transport resistance) and capacitance C (Fig. 3a) is linked to space charge region of the pn hetero-junction situated within the structure. Second combination of resistance R_3 (recombination resistance) and constant phase element (CPE) P_1 (Fig. 3a) should be associated with traps and defects in the structure [7, 9]. The parallel combination of R_2 and C is associated with semicircle in left part of Nyquist plot thus, the area of higher frequencies. Semicircle assigned to R_3 and P_1 (Fig. 3a) in lower frequency region is associated with internal charge transfer including accumulation P_1 and recombination R_3. As can be seen, the result of Xe exposure lead to change of electric behaviour of structure and necessarily the equivalent circuit had to be modified (Fig. 3b) as a result of disorders caused due to Xe irradiation of thin film structure.

IOP Conf. Series: Journal of Physics: Conf. Series **829** (2017) 012016 doi:10.1088/1742-6596/829/1/012016

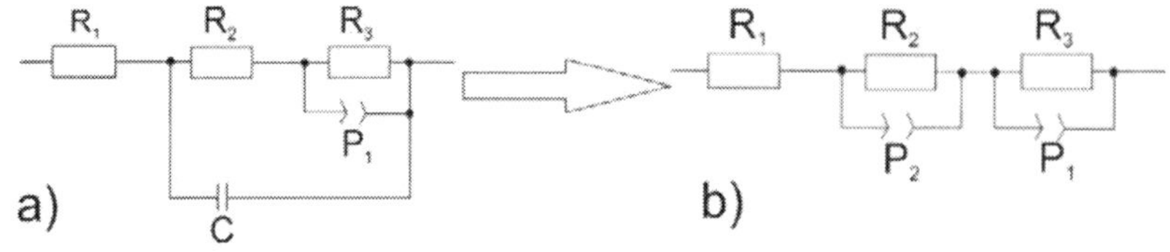

Figure 3. AC equivalent circuit before and after irradiation.

The behavior of respective equivalent circuit elements before and after irradiation is shown on Fig. 4 and 5. Fig. 4 shows the dependence of obtained parallel resistances R_2 and R_3 for the applied DC bias voltage before and after Xe irradiation. As can be seen in Fig.4a, parallel resistors of AC equivalent circuit have a standard behavior before the Xe irradiation. The resistance decreases with increasing DC voltage. Parallel resistance R_2 is strongly dependent on the applied bias mainly at forward voltages. As for R_2, it is similar before and after irradiation. Parallel resistance R_3 of modified circuit is practically independent on applied voltage after irradiation, shown in Fig. 4b, what indicates the deterioration of pn junction which we suppose to be governed more by ohmic process. Series resistance is an important parameter of photovoltaic structure governing the power output. But series resistance R_1 is not influenced by Xe bombardment in any remarkable way.

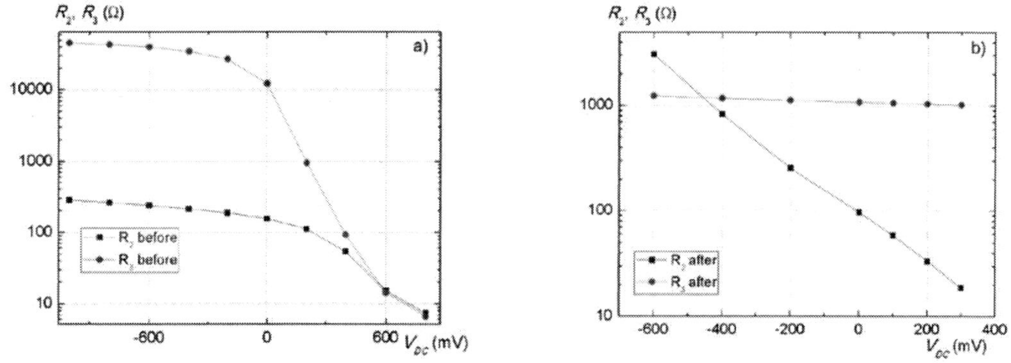

Figure 4. Parallel resistances of AC equivalent circuit before a) and after Xe irradiation b) for different applied DC bias in the dark at room temperature.

The dependences of numerically obtained values of capacitive elements at forward and reverse bias are presented in Fig. 5. The main contribution to parallel capacitance, under the reverse bias conditions is the depletion layer capacitance.

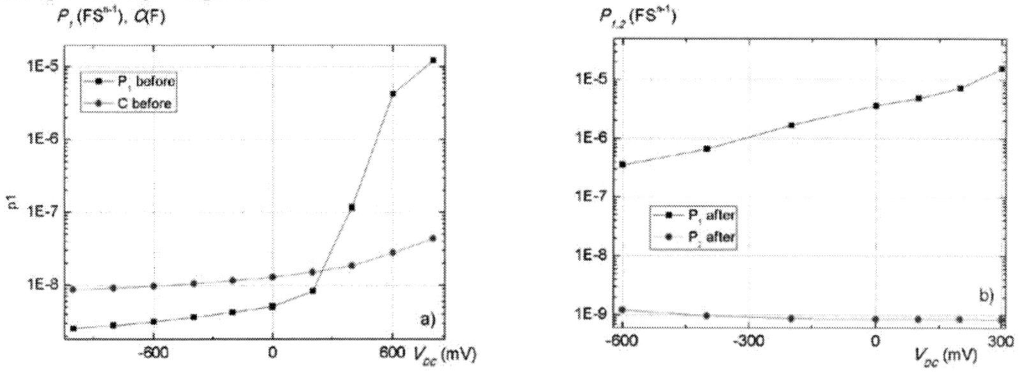

Figure 5. Parallel capacitances of AC equivalent circuit before and after Xe irradiation for different applied DC bias in the dark at room temperature.

At higher forward bias voltages, the depletion region associated with space charge region is disappearing (the capacitance C is not more dominant) therefore, the equivalent circuit evolves to a circuit of parallel R_3 and constant phase element P_1 in series with R_1. The increase of P_1 at positive biases before irradiation is related to the injection of minority charge carriers. Constant phase element was introduced into the equivalent AC circuit due to the properties of the structure (amorphous, disordered structure with the existence of structural defects). Introduction of constant phase element is in cases when the experimental data reveal a significant deviation from this simple R-C behavior, becoming apparent in a deformation and broadening of the impedance semicircle. A common phenomenological approach describes these broadened semicircle by a parallel connection of resistor R and a CPE.

Replacing the capacitor by the CPE but preserving the resistor in the circuit has the practical advantage that the DC conductivity can still be determined by directly fitting the parameter R [10].

Electrical impedance of CPE can be expressed as

$$Z_{CPE} = Y_0^{-1}(j\omega)^{-n}$$

where Y_0 represents CPE coefficient and n represents the character and an ideality of the distribution of relaxation times, ω is angular frequency [11].

The Xe bombardment led to change of conductive processes which is reflected in modified equivalent circuit. The circuit with two CPE´s after the Xe irradiation was proposed. In case of CPE capacitance P_1 increase of capacitance with increasing DC voltage was observed as shown in Fig. 5b while constant phase element P_2 is independent on applied voltage after Xe bombardment.

4. Conclusion

The solar heterojunction structure a-SiC/c-Si with used transparent conductive layer ITO which has been exposed to Xe irradiation was the object of research. Impact of the Xe radiation on electrical transport processes was monitored and evaluated by electrical impedance spectroscopy. AC equivalent circuit as the result of fitting the obtained impedance data was proposed and served to describe the electrical processes which result from radiation exposure. The most serious consequence of Xe exposure was significant increase of total resistance under forward bias which would lead to poor photovoltaic properties (not discussed at this place). The structural defects induced by Xe particles radiation affect also trapping mechanism what is illustrated by the influence of radiation on CPE included into equivalent circuit.

Acknowledgments

This work was supported by the Slovak Research and Development Agency under the contract No. APVV-0443-12 and by the Scientific Grant Agency of the Ministry of Education of the Slovak Republic and of the Slovak Academy of Sciences Project VEGA 1/0651/16. This paper is also the result of the project implementation: Finalizing of the National Centre for Research and Application of renewable Energy Sources, ITMS: 26240120028, supported by the R&D Operational Programme funded by the ERDF.

References

[1] Anderson D A and Spear W E 1977 *Philosophical Magazine B* **35** 113-131
[2] Strokan S B Ivanov A M Savkina N S, et al. 2004 *Semiconductors* Radiation resistance of transistor-and diode-type SiC detectors irradiated with 8-MeV protons **38** 807
[3] Masuko K, et al. 2014 *Photovoltaics* Achievement of more than 25% conversion efficiency with crystalline silicon heterojunction solar cell **6** 1433-5
[4] Janz S, Reber S, Glunz W 2006 *Proceedings of the 21st EUPVSEC* 660-663

[5] Polanský R, Pinkerová M, Bartůňková M and Prosr P 2013 *Journal of Electrical Engineering – Elektrotechnický časopis* Mechanical behaviour and thermal stability of EVA encapsulant material used in photovoltaic modules **64** 361 – 365

[6] Katoh Y, Sneada L L, Szlufarska I and Weber W J 2012 *Current Opinion in Solid State and Materials Science* Radiation effects in SiC for nuclear structural applications **16** 143–152

[7] Bisquert and Garcia-Belmonte G 2011 *J. Phys. Chem. Lett.* On voltage, photovoltage, and photocurrent in Bulk Heterojunction Organic Solar Cells **2** 1950-1964

[8] Huran J, Hrubčin L, Boháček P et al. 2015 *Proc. of Third International Conference on Radiation and Applications in Various Fields of Research (RAD 2015)* The effect of Xe and neutron irradiation on the properties of SiC and SiC(N) film prepared by PECVD technology, Montenegro

[9] Leever B J, Bailey C A, Marks T J, Hersam M C, and Durstock M F 2012 *Adv. Energy Mater.* In situ characterization of lifetime and morphology in operating bulk heterojunction organic photovoltaic devices by impedance spectroscopy **2** 120-128

[10] Shoar Abouzaria M R, Berkemeiera F, Schmitza G, Wilmer D 2009 *Solid State Ionics* On the physical interpretation of constant phase elements **180** Issues 14–16 922–927

[11] Hsu C H and Mansfeld F 2001 *Corrosion* Concerning the conversion of the constant phase element parameter Y0 into a capacitance **57** 747-748

IOP Conf. Series: Journal of Physics: Conf. Series **829** (2017) 012017 doi:10.1088/1742-6596/829/1/012017

SIKVAV peptide functionalized ultra-small gold nanoparticles for selective targeting of α6β1 integrin in hepatocellular carcinoma

M Roskamp[1], T Coulter[1], Y Ding[1], R Perrins[1], C Espinosa Garcia[1], A Pace[1], S Hale[1], A Robinson[1], P Williams[1], U Aguilera Peral[1], K Patel[1], D Palmer[1]

Midatech Pharma PLC, Abingdon, UK

dan@proficitpartners.com

Abstract. Ultra-small glycan-passivated gold nanoparticles of <2nm diameter were funtionalised with a short HS-EG(8)-COOH ligand. The nanoparticles were subsequently labelled, in a stoichiometrically controllable manner, with integrin-binding peptide SIKVAV and the maytansinoid cytotoxin DM4. *In vitro* assays showed significantly increased integrin-mediated uptake of SIKVAV labelled nanoparticles in HepG2 cells. SIKVAV targeted nanoparticle binding was shown to be outcompeted with free SIKVAV peptide, indicating target specific uptake. DM4 was passively attached to nanoparticles via sulfhydryl ligand exchange at the gold nanoparticle surface, which rendered them highly cytotoxic (IC50 ~1 x 10^{-9}M). In a rat model, pharmacokinetic studies showed that nanoparticle biodistribution was strongly altered by labelling with either peptide and DM4 moieties.

1. Introduction

Ultra-small gold nanoparticles (GNPs) have broad applicability in the field of drug delivery and tissue-specific drug delivery[1-3]. The term "ultra-small" is used to define nanoparticles with a core size of <2nm [4]. Numerous conflicting reports describe the toxicity of ultra-small GNPs[2,5-11]. Our own *in vivo* studies show across various species, that GNPs of this size with thiolated ligand shells can have negligible toxicity and immunogenicity. Differences in the ligand shell composition of GNP's may be a likely cause for differences in observed toxicity[7]. Ultra-small nanoparticles offer particularly high surface / volume ratio, which has important consequences for drug-loading and dose-efficiency. GNPs are also useful as drug-delivery vehicles, as their physicochemical properties may be fine-tuned to support various classes of therapeutic compounds[12-14].

Much of our research concerns the synthesis and functionalisation of glycan-passivated GNPs, which are afforded long-term stability and water-solubility by sulfhydryl-modified monosaccharide ligands such as 2′-thioethyl-α-D galactopyranoside[15,16]. Active targeting to specific tissues or cell types can be achieved by incorporating high-affinity species (e.g. peptides) within the ligand shell of a GNP construct[17-19]. For cancer therapy, such nanoparticle constructs are intended to enable the selective delivery of a pharmaceutical ingredient (often a cytotoxin) to tumour tissues, thus sparing healthy tissues and minimising off-target toxicities.

In this study, the peptide sequence (SIKVAV) derived from the basement membrane glycoprotein laminin α1 chain[20,21], was covalently attached via the N-terminus to carboxylated ultra-small GNPs. The sequence is known to bind to cell-surface integrins α3β1 and α6β1[22]. Importantly, the α6 integrin is known to be highly expressed by some cancer stem cells, notably those residing within the glioblastoma perivascular niche[23]. Similarly, hepatocytes are known to undergo unusual changes in expression of integrins upon carcinogenesis, including increased expression of laminin receptor $\alpha6\beta1$[24]. Other extracellular matrix-derived peptide sequences have been used to confer active-targeting functionality to a range of therapeutic nanoparticles. In a recent study, the widely studied integrin-binding sequence c(RGDfC) was covalently attached to ~5nm gold nanoparticles for the purposes of vascular targeting[18]. Additionally, the functionalization of larger (~300nm) carboxylate-functionalised nanoparticles with another laminin-derived peptide sequence (YIGSR) has previously been described[25] and shown to enable successful *in vivo* targeting to lung metastases.

The potent tubulin inhibitor tansine, which is derived from a naturally occurring ansamycin antibiotic, is an extremely potent cytotoxin that inhibits microtubule assembly and disrupts mitosis[26]. The structurally homologous mertansinoid DM4 has a terminal sulfhydryl function that can be used for conjugation as it is not related to activity. Despite high potency, DM4 has significant systemic side effects leading to poor tolerance and narrow therapeutic window. Here, we explore a straightforward approach to the synthesis of DM4-loaded GNPs via ligand exchange. This method was used to synthesise a novel GNP construct comprising SIKVAV peptide targeting ligands and DM4 cytotoxic ligands.

2. Results & Discussion

A previously characterised GNP construct was synthesised with a target ligand shell of 50% galactose-C2 thiol and 50% HS-EG(8)-COOH. The product (A) was characterised by DLS, ^{1}H-NMR, TEM and HPLC and was found to have, an average gold core size of ca. 1.8 nm and a hydrodynamic diameter of ca. 5 nm. Following purification, the GNP was reacted with an EDC/NHS mixture to convert carboxylic acid ligands to reactive NHS esters, then incubated with SIKVAV peptide. By varying the SIKVAV:GNP stoichiometry, GNPs could be produced with between 2 to 13 covalently coupled SIKVAV moieties per 200 gold atoms (Figure 1). For all further testing, a product GNP (B) ~~with~~ a ligand shell comprising [AuNP (α-Gal-C2)/(S-EG(8)COOH)/(S-EG(8)CONHSIKVAV)(11)] was chosen. This AuNP has an average of 8 SIKVAV peptides per 200 gold atoms.

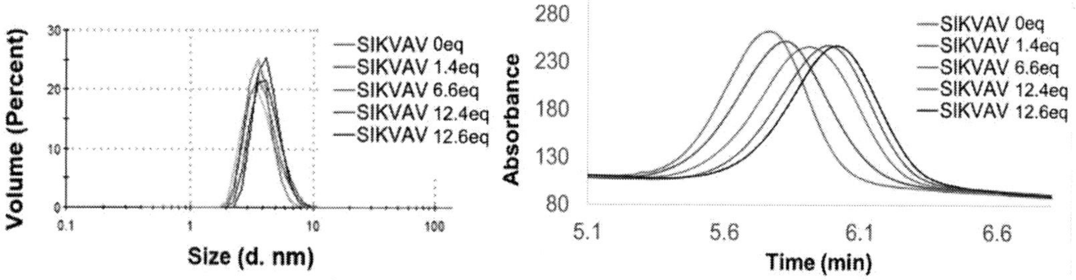

Figure 1 (A). DLS analysis. All samples were measured in PBS and showed an average hydrodynamic diameter of approximately 5nm. For each nanoparticle, the average of 3 measurements is shown. **(B). HPLC analysis** showed a shift of retention time for GNPs bearing varying equivalents of SIKVAV peptide.

To compare the effect of targeting ligand incorporation into our ultra-small GNPs, the ($\alpha6\beta1$) expressing cell line HEPG2[24] was exposed to non-targeted (GNP-**A**) and SIKVAV-targeted (**B**) particles. Silver enhancement staining revealed cell uptake to be ~2.5 fold higher in the SIKVAV-targeted GNPs (Figure 2a). To test the specificity of the observed increased GNP uptake, HEPG2 cells were pre-treated with free SIKVAV peptide before treatment with SIKVAV-labelled GNP **B**. Internalised gold content was measured by ICP-MS. These data revealed that pre-blocking with SIKVAV resulted in ~50% reduction in cell uptake (Figure 2b), achieving a similar level to that of the non-targeted GNP.

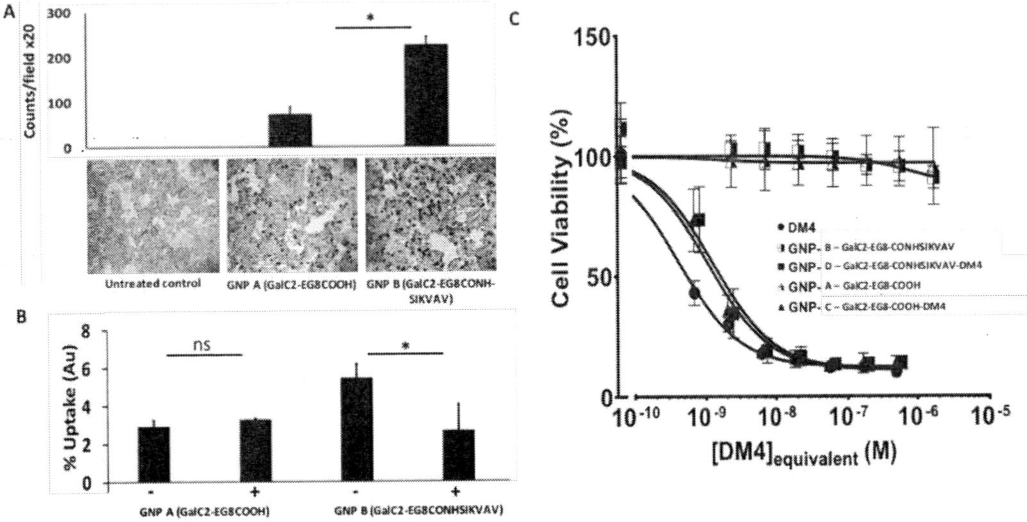

Figure 2A GNP-SIKVAV uptake in human hepatocellular carcinoma cells A) HepG2 cells were treated with 1x10[9] GNPs/cell for 1.5h at 37°C before fixation (4% PFA 10min; room temperature) and permeabilization (0.1%Triton X100 10min; RT). Cells were silver stained before scoring using a microscope at x20 magnification. B) HepG2 cells were treated with 1x10[9] GNPs/cell for 1.5h at 37°C to cells pre-incubated either in the absence (-) or presence (+) of free SIKVAV peptide (500nM). After exposure, cells were acid washed (0.2M Acetic acid/10.5M NaCl pH2.8: 10min at RT). Cell pellet gold content was measured by ICP-MS to quantify internalized GNPs. C) HepG2 cells were treated with GNPs functionalized with DM4 plus or minus SIKVAV targeting peptide for 48h before measurement of cell viability by MTT assay. Results were analyzed using a Student's *t*-test, (*), *p≤0.05*. Data are means ± standard deviation of at least three independent experiments.

2.1 Attachment of DM4

By treating glycan-passivated GNPs with another sulfhydryl ligand, ligand substitution is observed. The cytotoxin N2'-Deacetyl-N2'-(4-mercapto-4-methyl-1-oxopentyl)-maytansine (DM4) possesses a terminal sulfhydryl function. Here, we attempted to further functionalise GNPs with DM4 moieties by ligand exchange. The two GNP samples **A** & **B** were treated with 12 equivalents of DM4 per 200 gold atoms. After 3h, GNPs were isolated from the reaction mixture and carefully washed to remove traces of non-bound DM4. The non-peptide labelled GNP **A** underwent ligand substitution, yielding a new DM4 nanoparticle. By HPLC analysis, the product was estimated to have an average of 6 DM4 per 200 gold atoms bound to the GNP surface (using the model described above). The product nanoparticles were water soluble and showed no aggregation or precipitation in PBS. Similarly, a SIKVAV-labelled GNP **B** also gave effective DM4 substitution. HPLC analysis showed the isolated product GNP **(D)** to have a ligand shell with an estimated 8 DM4 per 200 gold atoms ~~were~~ substituted into the GNP ligand shell.

2.2 Toxicity studies

The toxicities of GNPs **A-D** were studied in HEPG2 cultures. Following treatment at numerous concentrations between 300 and 0.1ng/mL DM4 (GNPs A & B were tested at matching gold concentrations), for 72h, toxicity was determined by MTT assay. GNP constructs **A** & **B** showed no meaningful toxicity in HepG2 cultures at the highest concentration tested. However, the two DM4-labelled GNPs **C** & **D** were shown to be highly cytotoxic, with IC50 values of 1.0 x 10^{-9} and 1.3 x 10^{-9}M (DM4), respectively, (Figure 2C). From these results, we conclude that the attachment of DM4 (IC50 = 4.1 x 10^{-10}M) to ultra-small GNPs via the sulfhydryl function does not greatly diminish its cytotoxicity *in vitro*. The simple approach to drug-loading via competing sulfhydryl ligand exchange is potentially useful for the post-manufacture attachment of compounds with labile functional groups that are incompatible with the reducing, basic conditions of GNP synthesis.

2.3 In vivo biodistribution and PK

The GNP samples A-D were administered intravenously to male Wistar rats to elucidate the effects of peptide targeting & DM4 loading upon PK and biodistribution. Blood samples were collected at 1, 2, 4, 8 and 24h post injection. Attachment of the peptide sequence SIKVAV, which has a calculated LogP value of -0.75 is expected to introduce positive charge (from lysine residues) to the otherwise negatively charged GNP construct **A**. On the other hand, DM4 has a calculated LogP value of 5.72 and is expected to add significant hydrophobicity to the nanoparticle construct. Addition of SIKVAV reduced the zeta potential from -40 mV for **A** to -12 mV for **C** consistent with addition of a positively charged peptide. Incorporation of SIKVAV peptide or DM4 cytotoxin did affect both biodistribution and pharmacokinetics. Construct **A** showed the longest circulating half-life (Figure 3a). Construct **B**, labelled with an average of 7 SIKVAV peptide moieties per 200 gold atoms showed a similar elimination rate. Constructs loaded with DM4 moieties (**C** & **D**) showed a significantly increased elimination rate. From this preliminary study, it was not possible to derive meaningful PK parameters, hence further work is required to fully understand the impact of DM4 labelling on elimination rate and volume of distribution.

To more fully understand the fate of injected GNPs, tissue samples were harvested from liver, lungs, brain, eyes, faeces, femur, heart, kidney, large intestine, muscle, skin, plasma, small intestine, spleen, stomach, testes, urine and blood cells. For many of these tissues, ICP-MS analysis showed gold deposition to be extremely low and close to the limit of detection. Hence, we concentrated our analysis on the samples containing the bulk of the dose, i.e. liver, kidney, faeces, plasma and urine. Construct **A** showed minimal tissue deposition, with just 9% of dose recovered in the kidney and over 65% recovered in urine. Interestingly, the incorporation of SIKVAV peptide (**B**) caused minimal change in distribution to the liver and plasma, but a marked increase in kidney deposition (45% of dose) and a decreased elimination in urine (35%). GNPs loaded with DM4 (**C**) showed a different pattern of distribution, in which liver uptake was double that of **A**, whilst plasma levels were the lowest detected at 0.6% of dose (corresponding with pharmacokinetic profile above). Again, deposition to kidney increased to 19.5%, and urine content decreased to 22% of the given dose. Finally, construct **D**, bearing both SIKVAV and DM4 moieties, showed the highest uptake in kidney (60%) and a large reduction in urine content (27% of dose) (Figure 3b). Accumulation of peptide GNP's in the kidney has been observed in previous studies (unpublisehed observation).

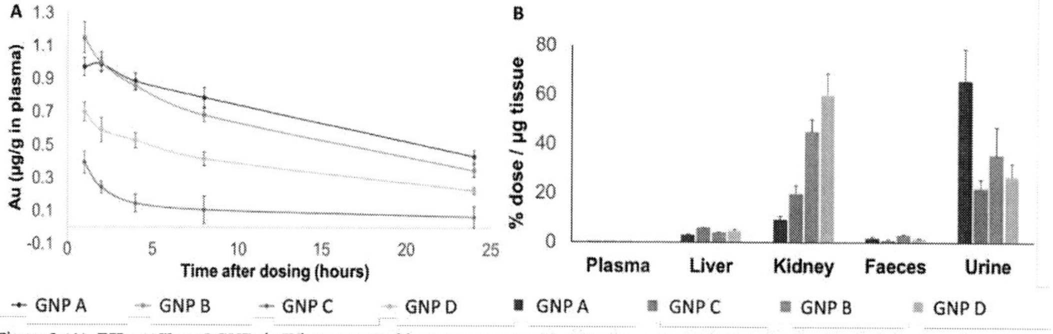

Figure 3 (A). **PK profiles of GNPs** in Wistar rats: Gold-content measured in blood samples following single tail vein injection of GNPs A-D (300µg/kg dose). Concentrations measured by ICP-MS (**B**). **Biodistribution of GNPs** 24h post dosing. Gold-content found in major organs & tissues following tail vein injection of GNPs A-D. Concentrations measured by ICP-MS. GNP A = $GalC_2$-EG_8-COOH; B = $GalC_2$-EG_8-CONHSIKVAV; C = $GalC_2$-EG_8-COOH-DM4; D = $GalC_2$-EG_8-CONHSIKVAV-DM4

3. Materials & Methods

3.1 Chemicals & materials

Dimethyl sulfoxide, Hydrogen tetrachloroaurate (III) trihydrate ($HAuCl_4.3H_2O$), EDC (N-(3-Dimethylamino-propyl)-N□-ethylcarbodiimide hydrochloride); DMEM medium, Ethanol (absolute); L-Glutamine, Methylthiazolyl-diphenyl-tetrazolium bromide; foetal calf serum; NHS (N-Hydroxy-succinimide); Non-essential amino acids; ProteoSilver™ Silver Stain Kit; Sodium borohydride; Sodium hydroxide and Sodium pyruvate were all obtained from Sigma-Aldrich, UK. N2'-Deacetyl-N2'-(4-mercapto-4-methyl-1-oxopentyl)-maytansine, referred to as DM4, was purchased from Carbosynth, UK. Di-2′-thioethyl-α-D galactopyranoside disulfide was purchased from Galchimia, A Coruña, Spain. SIKVAV peptide was purchased from Bachem, Switzerland. 1-mercapto-3,6,9,12,15,18,21,24-octaoxaheptacosan-27-oic acid (HS-$EG_{(8)}$COOH) was purchased from Iris

Biotech GmbH, Germany. Amicon Ultra-15 centrifugal filter units, MWCO=10K were purchased from Millipore Sigma, UK. Deionized water was obtained from an Elga Purelab Option-Q (DV25) (electrical resistivity 18.2MΩ).
Analytical methods

Analytical methods used for physico-chemical characterisation of GNPs are fully described in the **supplementary information**. Briefly, UV-Vis absorption spectra were measured using a BMG Labtech SPECTROstar Nano microwell plate spectrophotometer. Dynamic light scattering was measured in triplicate using a Zetasizer Nano ZSP (Malvern Instruments Ltd, UK), at 25°C in distilled water. Mean volume-weighted hydrodynamic diameter of the particles was used as a measure of GNP size. GNP-bound ligands were analysed by ^{1}H-NMR after dissolution of gold cores by treatment with KCN/KOH (0.3M/0.1M, respectively in D_2O). HPLC analysis was performed using an Agilent 1260 Infinity system with an Ascentis Express Peptide C-18 column (4.6 × 100 mm, 2.7µm; Sigma; 35°C column temperature; flow rate 1ml/min). Ligand shell composition and DM4 loading were analysed by releasing the ligands from GNPs via incubation in 100mM Dithiothreitol for 3h.

3.2 Nanoparticle Synthesis
Carboxyl-functionalized ultra-small GNPs were synthesised at a 10mg scale via a modified Brust-Schiffrin method[28]. Aqueous solutions of 1-mercapto-3,6,9,12,15,18,21,24-octaoxaheptacosan-27-oic acid (HS-$EG_{(8)}$-COOH) (7.5mL, 20mM) and di-2´-thioethyl-α-D galactopyranoside disulfide (2.5mL, 20mM) were mixed in a round bottomed flask. With continuous stirring, an aqueous solution of $HAuCl_4$ (88µL, 333mM) was added. The pH was adjusted to 11.6 by adding 2M NaOH solution. A solution of sodium borohydride (1mL, 1.0M) was then added as quickly as possible. The resulting dark purple/black solution was stirred for a further 1 hour before purification by ultra-centrifugation and repeated washing in distilled water in an Amicon Ultra-15 (MWCO 10kDa) tube. Characterisation by DLS, HPLC and 1H-NMR revealed the product GNP (**A**) had an average diameter of 4nm and an average ligand shell comprising 40% galactose-C_2 and 60% HS-$EG_{(8)}$-COOH, which is expressed as [AuNP(α-Gal-C_2)$_{(40)}$ / (S-$EG_{(8)}$COOH)$_{(60)}$].

3.3 Conjugation of SIKVAV peptide
GNP construct **A** (1.0mL, 3g/L Au) was activated with a mixture of EDC (55eq. per gold) and NHS (110eq. per gold) in DMSO for 2 hours. The activated GNPs were diluted in deionised water and centrifuged in an Amicon Ultra-15 tube to remove unreacted reagents. A solution of SIKVAV in 40mM HEPES buffer (pH7.8) was prepared (2.0mL, 0.263mM), to which the activated carboxyl GNPs were added via pipette. The mixture was stirred at room temperature overnight. The product was purified by ultra-centrifugation and repeated washing in deionised water in an Amicon Ultra-15 (MWCO 10kDa) tube.

3.4 DM4 Ligand exchange reactions
DM4 in absolute ethanol (500µL, 2.36mM) was carefully pipetted into a glass vial. An aqueous solution of the GNP [(α-Gal-C_2)$_{(40)}$ / (S-$EG_{(8)}$COOH)$_{(60)}$] was added (500µL, 4g/L Au). The mixture was stirred for 3h at RT. After this time, the nanoparticle solution was transferred to an Amicon Ultra-15 tube and diluted with 7mL 50% ethanol solution. The product was then purified by repeated centrifugation at 5000rpm, during which the solvent was gradually changed to water. A small, white solid pellet (unreacted DM4) was removed by careful pipetting. HPLC analysis showed the extent of ligand exchange and DM4 loading per nanoparticle. Nanoparticle-bound DM4 was quantified by comparison with a standard curve. Using an analogous method, SIKVAV-labelled nanoparticles were treated with free DM4 for 3h to allow passive ligand exchange, yielding a brown solution that was analysed for DM4 content by HPLC.

3.5 Biological Methods
HEPG2 cells (human hepatocellular carcinoma, ATTC) were cultured in DMEM supplemented with 10% (v/v) FBS, 1% (v/v), L-Glutamine, non-essential amino acid and sodium pyruvate, at 37°C & 5% CO_2 atmosphere. Cells were sub-cultured regularly and used at passage +5-10. MTT cytotoxicity assays were performed using a standard procedure, outlined in **supplementary information**.
Adherent cell cultures were prepared in 96 well-plates and incubated overnight. Growth medium was replaced with solutions of test GNPs in low FBS medium (2%) at a concentration of 1.0×10^9 GNPs per cell (20µL volume). Cultures were incubated for 90 min at 37°C, before being washed with warm PBS. Cells were fixed in 4% PFA solution (50µL per well) and incubated at RT for 10 min. Cell cultures were permeabilised by addition

of Triton X-100 in PBS (0.1%, 50μL per well) for 10 min, then carefully washed with PBS. Working quickly, cell cultures were treated with the pre-mixed silver stain solutions at RT (100μL each) and imaged by light microscopy.

GNP uptake was also quantified (gold content) by ICP-MS. HEPG2 cells were seeded in 6 well-plates at a density of $1.2*10^6$ cells in 2mL growth medium and allowed to grow to 95% confluency. The medium was replaced with a solution of $1.0x10^8$ GNPs per cell in low-serum medium (400μL, 2% FBS). Treated cultures were incubated for 3 hours, then washed with PBS (1mL). Cells were scraped from well-plates, transferred to 2mL tubes and centrifuged at 200 x g for 5 min before being washed with a chilled solution of acetic acid (0.2M in 10.5M NaCl, pH2.8; 1mL; 4°C). After 6 min, samples were centrifuged at 200 x g for 5 min then incubated at RT for 10 min. Pellets were washed a further two times in PBS (1mL) and then frozen at -20°C. Before analysis, the pellets were digested in a mixture of nitric acid, hydrochloric acid and hydrogen peroxide (1.6mL, 0.5mL, 0.04mL, respectively) and diluted to 10mL volume in deionized water. The ensuing solutions were analysed using an Agilent 7700X ICP mass spectrometer alongside appropriate calibration standards.

For PK and biodistribution studies, male Wistar rats (7-8 weeks old, $n = 4$) were injected via the tail vein with a single dose of GNP sample at 0.3 mg(Au)/kg gold. Blood plasma samples (250μL) were collected at 0, 1, 2, 4, 8 and 24h post injection. All major organs & tissues were harvested and wet-weighed. Plasma & tissue samples were digested in Teflon microwave vessels using a CEM Discover Microwave digestion system. Gold concentrations were then measured by ICP-MS with appropriate calibration standards and expressed as a percentage of injected dose per gram of tissue.

4. Conclusion

Over-expression of integrins such as α6β1 is reported to be a characteristic of some cancer stem cells, glioblastomas and hepatocellular carcinomas. Hence, increased specific cytotoxicity and diminished systemic toxicity could be expected of a therapeutic GNP bearing SIKVAV peptides that bind with high affinity to α6β1 integrins. SIKVAV was covalently attached to glycan/$EG_{(8)}$-COOH ultra-small GNPs via EDC/NHS coupling to the carboxylated ligand HS-$EG_{(8)}$-COOH. The product GNPs were further derivatised with the microtubule inhibitor DM4, yielding stable, water-soluble constructs. *In vitro* experiments showed SIKVAV labelling produced a significant increase in nanoparticle uptake in HEPG2 cells. Similarly, *in vitro* toxicity studies showed DM4-loaded GNPs to be highly toxic to this cell line. We observed a marked increase in nanoparticle elimination rate upon attachment of DM4 moieties, which is worthy of further study. We propose further development of the GNP construct, to extend plasma half-life and to improve target binding efficacy by modifying the targeting peptide / drug/ligand ratio.

References

[1] Kumar, Anil, Xu Zhang, and Xing-Jie Liang. "Gold Nanoparticles: Emerging Paradigm for Targeted Drug Delivery System." Biotechnology Advances 31.5 (2013): 593-606. Web.
[2] Ajnai, Giimel, Amy Chiu, Tzuchun Kan, Chun-Chia Cheng, Teh-Hua Tsai, and Jungshan Chang. "Trends of Gold Nanoparticle-based Drug Delivery System in Cancer Therapy." Journal of Experimental & Clinical Medicine 6.6 (2014): 172-78. Web.
[3] Male, D., R. Gromnicova, and C. Quaid. "Chapter 5: Gold Nanoparticles for Imaging and Drug Transport to the CNS." Nanotechnology and the Brain. Vol. 130. N.p.: Elsevier, 2016. 155-98. Print. International Review of Neurobiology.
[4] Zarschler, Kristof, Louise Rocks, Nadia Licciardello, Luca Boselli, Ester Polo, Karina Garcia Pombo, Luisa Cola De, Holger Stephan, and Kenneth Dawson A. "Ultrasmall Inorganic Nanoparticles: State-of-the-art and Perspectives for Biomedical Applications." Nanomedicine: Nanotechnology, Biology and Medicine 12.6 (2016): 1663-701. Web.
[5] Carnovale, Catherine, Gary Bryant, Ravi Shukla, and Vipul Bansal. "Size, Shape and Surface Chemistry of Nano-gold Dictate Its Cellular Interactions, Uptake and Toxicity." Progress in Materials Science 83 (2016): 152-90. Web.
[6] Connor, Ellen E., Judith Mwamuka, Anand Gole, Catherine Murphy J., and Michael Wyatt D. "Gold Nanoparticles Are Taken Up by Human Cells but Do Not Cause Acute Cytotoxicity." Small 1.3 (2005): 325-27. Web.

[7] Pan, Yu, Sabine Neuss, Annika Leifert, Monika Fischler, Fei Wen, Ulrich Simon, Günter Schmid, Wolfgang Brandau, and Willi Jahnen-Dechent. "Size-Dependent Cytotoxicity of Gold Nanoparticles." Small 3.11 (2007): 1941-949. Web.

[8] Schmid, Günter. "The Relevance of Shape and Size of Au55 Clusters." Chemical Society Reviews Chem. Soc. Rev. 37.9 (2008): 1909. Web.

[9] Chen, Yu-Shiun, Yao-Ching Hung, Ian Liau, and G. Huang Steve. "Assessment of the *In Vivo* Toxicity of Gold Nanoparticles." Nanoscale Res Lett Nanoscale Research Letters 4.8 (2009): 858-64. Web.

[10] Alkilany, Alaaldin M., and Catherine Murphy J. "Toxicity and Cellular Uptake of Gold Nanoparticles: What We Have Learned so Far?" Journal of Nanoparticle Research J Nanopart Res 12.7 (2010): 2313-333. Web.

[11] Fraga, Sónia, Ana Brandão, Maria Soares Elisa, Tiago Morais, José Duarte Alberto, Laura Pereira, Leonor Soares, Cristina Neves, Eulália Pereira, Maria Bastos De Lourdes, and Helena Carmo. "Short- and Long-term Distribution and Toxicity of Gold Nanoparticles in the Rat after a Single-dose Intravenous Administration." Nanomedicine: Nanotechnology, Biology and Medicine 10.8 (2014): 1757-766. Web.

[12] Amirthalingam, Ezhil, Mafalda Rodrigues, Lucía Casal-Dujat, Ana Calpena C., David Amabilino B., David Ramos-López, and Lluïsa Pérez-García. "Macrocyclic Imidazolium-based Amphiphiles for the Synthesis of Gold Nanoparticles and Delivery of Anionic Drugs." Journal of Colloid and Interface Science 437 (2015): 132-39. Web.

[13] Pooja, Deep, Sravani Panyaram, Hitesh Kulhari, Bharathi Reddy, Shyam Rachamalla S., and Ramakrishna Sistla. "Natural Polysaccharide Functionalized Gold Nanoparticles as Biocompatible Drug Delivery Carrier." International Journal of Biological Macromolecules 80 (2015): 48-56. Web.

[14] Vigderman, Leonid, and Eugene Zubarev R. "Therapeutic Platforms Based on Gold Nanoparticles and Their Covalent Conjugates with Drug Molecules." Advanced Drug Delivery Reviews 65.5 (2013): 663-76. Web.

[15] Ojeda, Rafael, Jose Paz Luis De, Africa Barrientos G., Manuel Martín-Lomas, and Soledad Penadés. "Preparation of Multifunctional Glyconanoparticles as a Platform for Potential Carbohydrate-based Anticancer Vaccines." Carbohydrate Research 342.3-4 (2007): 448-59. Web.

[16] Paz, José-Luis De, Rafael Ojeda, África Barrientos G., Soledad Penadés, and Manuel Martín-Lomas. "Synthesis of a Ley Neoglycoconjugate and Ley-functionalized Gold Glyconanoparticles." Tetrahedron: Asymmetry 16.1 (2005): 149-58. Web

[17] Galbiati, Elisabetta, Luca Gambini, Viola Civitarese, Michela Bellini, Dario Ambrosini, Raffaele Allevi, Svetlana Avvakumova, Sergio Romeo, and Davide Prosperi. ""Blind" Targeting in Action: From Phage Display to Breast Cancer Cell Targeting with Peptide-gold Nanoconjugates." Pharmacological Research 111 (2016): 155-62. Web.

[18] Peiris, Pubudu M., Partha Deb, Elizabeth Doolittle, Gilad Doron, Amy Goldberg, Priya Govender, Shruti Shah, Swetha Rao, Sarah Carbone, Thomas Cotey, Meilyn Sylvestre, Sohaj Singh, William Schiemann P., Zhenghong Lee, and Efstathios Karathanasis. "Vascular Targeting of a Gold Nanoparticle to Breast Cancer Metastasis." Journal of Pharmaceutical Sciences 104.8 (2015): 2600-610. Web.

[19] Han, Liang, Qiang Cai, Daofeng Tian, Derek Kong K., Xingchun Gou, Zeming Chen, Stephen Strittmatter M., Zuoheng Wang, Kevin Sheth N., and Jiangbing Zhou. "Targeted Drug Delivery to Ischemic Stroke via Chlorotoxin-anchored, Lexiscan-loaded Nanoparticles." Nanomedicine: Nanotechnology, Biology and Medicine 12.7 (2016): 1833-842. Web.

[20] Nomizu, Motoyoshi, Benjamin Weeks S., Christi Weston A., Woo Kim Hoo, Hynda Kleinman K., and Yoshihiko Yamada. "Structure-activity Study of a Laminin α1 Chain Active Peptide Segment Ile-Lys-Val-Ala-Val (IKVAV)." FEBS Letters 365.2-3 (1995): 227-31. Web.

[21] Okur, Aysu Ceren, Pelin Erkoc, and Seda Kizilel. "Targeting Cancer Cells via Tumor-homing Peptide CREKA Functional PEG Nanoparticles." Colloids and Surfaces B: Biointerfaces 147 (2016): 191-200. Web.

[22] Kikkawa, Yamato, Kentaro Hozumi, Fumihiko Katagiri, Motoyoshi Nomizu, Hynda Kleinman K., and Jennifer Koblinski E. "Laminin-111-derived Peptides and Cancer." Cell Adhesion & Migration 7.1 (2013): 150-59. Web.

[23] Lathia, Justin D., Joseph Gallagher, John Heddleston M., Jialiang Wang, Christine Eyler E., Jennifer Macswords, Qiulian Wu, Amit Vasanji, Roger Mclendon E., Anita Hjelmeland B., and Jeremy Rich N. "Integrin Alpha 6 Regulates Glioblastoma Stem Cells." Cell Stem Cell 6.5 (2010): 421-32. Web.

[24] Nejjari, Mimoun, Zakia Hafdi, Jerome Dumortier, Annie-France Bringuier, Gerard Feldmann, and Jean-Yves Scoazec. "α6β1 Integrin Expression in Hepatocarcinoma Cells: Regulation and Role in Cell Adhesion and Migration." International Journal of Cancer Int. J. Cancer 83.4 (1999): 518-25. Web.

[25] Sarfati, Gadi, Tal Dvir, Moshe Elkabets, Ron Apte N., and Smadar Cohen. "Targeting of Polymeric Nanoparticles to Lung Metastases by Surface-attachment of YIGSR Peptide from Laminin." Biomaterials 32.1 (2011): 152-61. Web.

[26] Poon, Kirsten Achilles, Kelly Flagella, Joseph Beyer, Jay Tibbitts, Surinder Kaur, Ola Saad, Joo-Hee Yi, Sandhya Girish, Noel Dybdal, and Theresa Reynolds. "Preclinical Safety Profile of Trastuzumab Emtansine (T-DM1): Mechanism of Action of Its Cytotoxic Component Retained with Improved Tolerability." Toxicology and Applied Pharmacology 273.2 (2013): 298-313. Web.

[27] Hostetler, Michael J., Julia E. Wingate, Chuan-Jian Zhong, Jay E. Harris, Richard W. Vachet, Michael R. Clark, J. David Londono, Stephen J. Green, Jennifer J. Stokes, George D. Wignall, Gary L. Glish, Marc D. Porter, Neal D. Evans, and Royce W. Murray. "Alkanethiolate Gold Cluster Molecules with Core Diameters from 1.5 to 5.2 Nm:Â Core and Monolayer Properties as a Function of Core Size." Langmuir 14.1 (1998): 17-30. Web.

[28] Brust, Mathias, Merryl Walker, Donald Bethell, David Schiffrin J., and Robin Whyman. "Synthesis of Thiol-derivatised Gold Nanoparticles in a Two-phase Liquid–Liquid System." J. Chem. Soc., Chem. Commun. 0.7 (1994): 801-02. Web.

IOP Conf. Series: Journal of Physics: Conf. Series **829** (2017) 012018 doi:10.1088/1742-6596/829/1/012018

Fully developed nanofluid mixed convection flow in a vertical channel

Fahad G. Al-Amri

Department of Basic Engineering, College of Engineering, University of Dammam, P.O. Box 1982, Dammam 31441, Saudi Arabia

fgalamri@uod.edu.sa

Abstract. A closed form analytical solution of laminar mixed convection heat transfer of a nanofluid between two vertical parallel plates, accounting for the effects Brownian motion and thermophoresis, is presented for the fully developed region under thermal boundary conditions of the first and fourth kinds. Four kinds of nano-sized solid particles with varied thermophysical properties suspended in water are considered. Closed form analytical expressions of velocity and temperature fields, pressure gradients, nanoparticle concentration profiles, and Nusslet numbers are illustrated. Effects of the controlling parameters, namely the buoyancy parameter, thermal conductivity, solid/fluid ratio, and volume fraction, on the hydrodynamic and heat transfer parameters such as pressure gradient and Nusslet number are discussed in detail. It is found that the Nusslet number increases with increases in the buoyancy parameter and volume fraction. However, the pressure drop is found to increase with volume fraction and decrease with the buoyancy parameter. In addition, for upward mixed convection flow, the pressure drop attributed to the addition of nano-sized solid particles into the base fluid can be overcome by the buoyancy forces. The critical values of the buoyancy parameter– where the buoyancy forces balance the viscous forces–are obtained and presented.

1. Introduction

Nanofluids are promising cooling mediums that are expected to replace conventional media in many engineering applications that require high-efficiency cooling systems. They can be produced by adding a small amount of tiny-sized solid particles to a conventional fluid such as water or oil. Because the thermal conductivities of the solid particles are high compared to the base fluid, the thermal conductivity of the mixture increases, leading to heat transfer enhancement. However, experimental work showed that this enhancement is much higher than the expected thermal conductivity increment [1]. Two slip mechanisms producing a relative velocity between the nanoparticle and the base fluid have been proven to be the most significant sources behind the anomalous heat transfer in nanofluids: Brownian motion and thermophoresis [2].

Several researchers studied the effects of Brownian diffusion and thermophoresis on nanofluid heat transfer flow inside channels by employing the two-component non-homogeneous model for convection transport in nanofluids, developed by Buongiorno [2]. Grosan and Pop [3] and di Schio et al. [4] investigated the problems of mixed and forced convection flows between two parallel plates, respectively. Yang et al. [5] provided a theoretical explanation for the anomalous heat transfer in nanofluids. Li and Nakayama [6] studied the effect of temperature dependency of nanofluid thermophysical properties on forced convection heat transfer enhancement. In all cases, an obvious increment in the heat transfer of nanofluids was found.

In this paper, the problem of laminar mixed convection heat transfer of a nanofluid between two vertical parallel plates, taking into account the Brownian motion and thermophoresis effects, is investigated in the fully developed region under the thermal boundary conditions of the first and fourth kinds. Based on a literature survey, the problem has not yet been solved, and such a study is important to understanding how nanoparticle migration impacts heat transfer inside channels. The presented closed form solutions can also be used to validate future work conducted in the developing region.

2. Governing equations and solution

A fully developed mixed convection nanofluid flow between two vertical plates under the first and fourth kinds of boundary conditions is considered. At low volume fraction, the nanoparticles and the base fluid form a dilute mixture and are in thermal equilibrium. The thermophysical properties of the nanoparticles and the base fluid (water) used in this study can be found elsewhere in the literature. The viscosity, thermal conductivity, and density of the nanofluid are hereby evaluated using the following relationships.

$$\mu_{nf} = \frac{\mu_{bf}}{(1-\phi_o)^{2.5}}$$

$$\rho_{nf} = (1-\phi_o)\rho_{bf} + \phi_o \rho_p$$

$$K_{nf} = K_{bf}\left[\frac{K_p + 2K_{bf} - 2\phi_0 (K_{bf} - K_p)}{K_p + 2K_{bf} + \phi_o (K_{bf} - K_p)}\right]$$

For laminar, steady flow in the fully developed region, the conversion equations that govern the present problem can be written in dimensionless forms.

Momentum equation:

$$\frac{d^2U}{dY^2} - (1-\phi_0)^{2.5}\frac{dP}{dZ} + \frac{Gr}{Re}\left[(1-\phi_0)^{3.5}\theta - \frac{\gamma}{2+Kr}\phi_0(1-\phi_0)^{2.5} N_{BT}(\Phi - 1)\right] = 0 \tag{1}$$

Energy equation:

$$\frac{\partial^2 \theta}{\partial Y^2} = 0 \tag{2}$$

Continuity equation for the nanoparticles:

$$N_{BT}\frac{\partial^2 \Phi}{\partial Y^2} + \frac{\partial^2 \theta}{\partial Y^2} = 0 \tag{3}$$

Mass flux concentration of the nanoparticles:

$$\int_0^1 \Phi \, dy = 1 \tag{4}$$

Integral form of the nanofluid continuity equation:

$$\int_0^1 U \, dY = 1 \tag{5}$$

All these equations are subject to boundary conditions, which are U(0)=0, U(1)=0, $\theta(0)=1$, and $\theta(1)=0$ for the first kind of boundary condition and $\left.\frac{\partial\theta}{\partial Y}\right|_{Y=0} = -\frac{k_{bf}}{k_{nf}}$ and $\theta(1)=0$ for the fourth kind of boundary condition. The walls are impermeable; hence additional boundary conditions can be written as $N_{BT}\frac{\partial\theta}{\partial Y}|_{Y=0} + \frac{\partial\theta}{\partial Y}|_{Y=0} = 0$ and $N_{BT}\frac{\partial\theta}{\partial Y}|_{Y=1} + \frac{\partial\theta}{\partial Y}|_{Y=1} = 0$. These governing equations, subject to their boundary conditions, are solved analytically, and closed form expressions are obtained.

For the first kind of boundary condition:

$$\theta = -Y+1 \tag{6}$$

$$\frac{dP}{dZ} = \frac{-12+0.5*Gr/Re*(1-\phi_0)^{3.5}}{(1-\phi_0)^{2.5}} \tag{7}$$

$$Nu = \frac{51.428*K_{nf}/K_{bf}}{25.71-0.0713*Gr/Re((1-\phi_0)^{3.5}+\gamma/(2+k_p/k_{bf})\phi_0(1-\phi_0)^{2.5})} \tag{8}$$

For the fourth kind of boundary condition:

$$\theta = -k_{bf}/k_{nf}\,(-Y+1) \tag{9}$$

$$\frac{dP}{dZ} = \frac{-12+0.5*K_{bf}/K_{nf}*Gr/Re*(1-\phi_0)^{3.5}}{(1-\phi_0)^{2.5}} \tag{10}$$

$$Nu = \frac{51.428*K_{nf}/K_{bf}}{25.71-0.0713*K_{bf}/K_{nf}*Gr/Re((1-\phi_0)^{3.5}+\gamma/(2+k_p/k_{bf})\phi_0(1-\phi_0)^{2.5})} \tag{11}$$

The dimensionless velocity for both kinds is:

$$U=Y*(c_1*(-0.25+0.25*Y)+c_2*dP/dZ*(-0.5+0.5*Y)+c_3*(1/3+Y*(-1/2+1/6*Y))) \tag{12}$$

Where $c_1=\frac{Gr}{Re}\left[(1-\phi_0)^{3.5}\theta-\frac{\gamma}{2+Kr}\phi_0(1-\phi_0)^{2.5}N_{BT}(\Phi-1)\right]$, $c_2=(1-\phi_0)^{2.5}$, and

$c_3=\frac{Gr}{Re}\left[-\frac{\gamma}{2+Kr}\phi_0(1-\phi_0)^{2.5}N_{BT}(\Phi-1)\right]$.

An inspection of Equations 7 and 10 reveals that the pressure drop inside the channel increases with the presence of nanoparticles, and the higher the volume fraction, the greater the effect. It also reveals that the buoyancy reduces the negativity of the pressure gradient buoyancy-aided flow. Thus, there is a value of Gr/Re that would reduce the pressure gradient to its value for pure fluid (i.e., when no solid particles are added to the fluid). The value can be obtained by equating the pressure given by Equations 7 and 10 to -12 and is given as:

$$\left.\frac{Gr}{Re}\right|_{cr} = 2*(-12+12*(1-\phi_0)^{-2.5})*(1-\phi_0)^{-1} *w \tag{13}$$

Where w=1 for the first kind of boundary condition, and $w=k_{nf}/k_{bf}$ for the fourth kind of boundary condition.

3. Results and Discussion

Figure 1 shows the variation in pressure gradient with the volume fraction for three values of the buoyancy parameter. It is clear from the figure that the pressure drop inside the channel increases when nanoparticles are added to the fluid. The higher the volume fraction, the greater the effect. In contrast, the buoyancy force helps to reduce the pressure drop in the channel, also shown in the figure. Figure 2 shows the pressure gradient variation with volume fraction for three values of the solid/fluid thermal conductivity ratio for the fourth kind of boundary condition. Clearly, the pressure drop decreases as the thermal conductivity ratio decreases. Figures 3 and 4 show the variation of Nu with volume fraction for various values of Gr/Re and k_p/k_f, respectively. As expected, Nu increases with volume fraction and the buoyancy parameter. However, it decreases with the solid/fluid conductivity ratio, and this is attributed to the migration enhancement that results from the increment in thermophoresis.

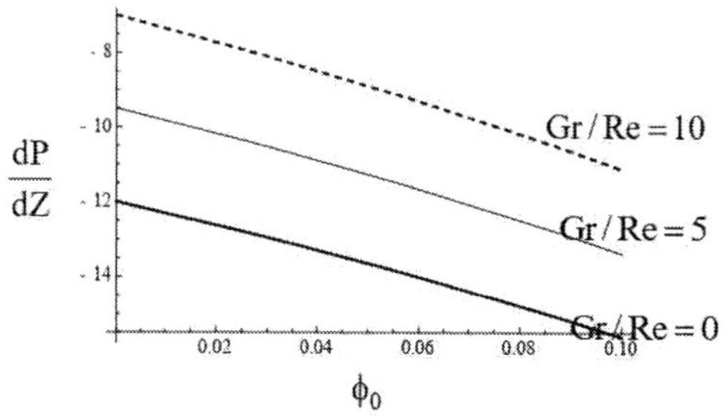

Figure 1. Pressure gradient as a function of volume fraction for the first kind of boundary condition.

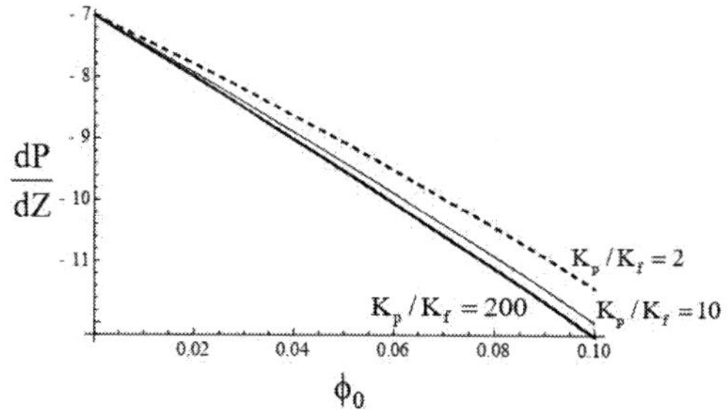

Figure 2. Pressuregradient as a function of volume fraction for the fourth kind of boundary condition (Gr/Re=10).

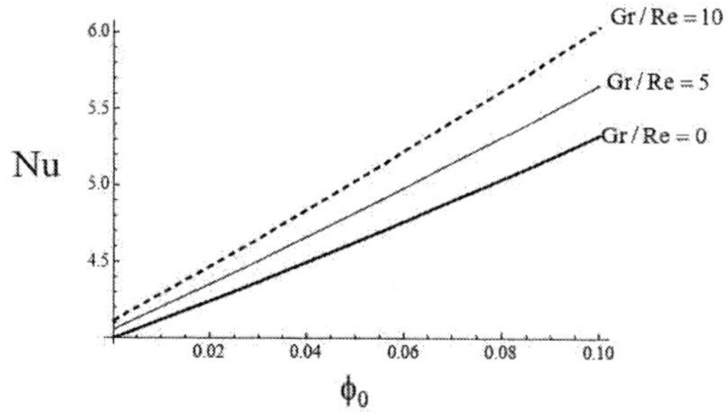

Figure 3. Nu as a function of volume fraction for the first kind of boundary condition (Kr=650).

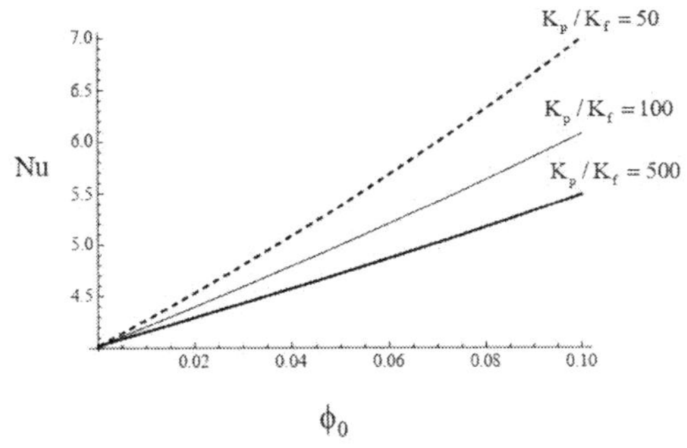

Figure 4. Nu as a function of volume fraction for the first kind of boundary condition (Gr/Re=2).

Figures 5 and 6 represent the pressure gradient and the Nusslet number, respectively, as functions of volume fraction for four kinds of nanoparticles. They show that SiO_2 gives the minimum pressure drop inside the channel, while Al_2O_3 gives the maximum heat transfer rate.

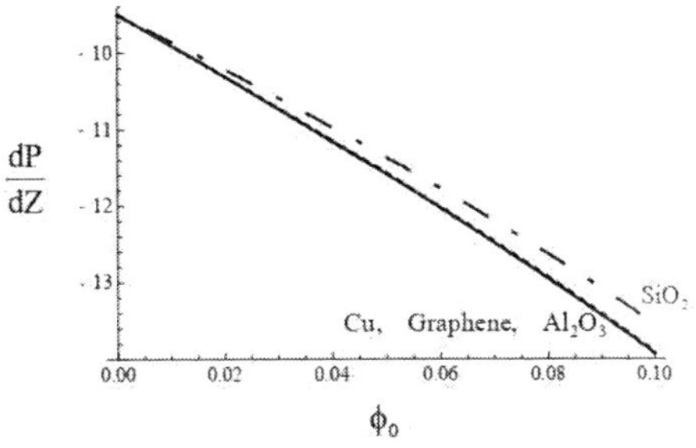

Figure 5. Pressure gradient as a function of volume fraction for the fourth kind of boundary condition (Gr/Re=5).

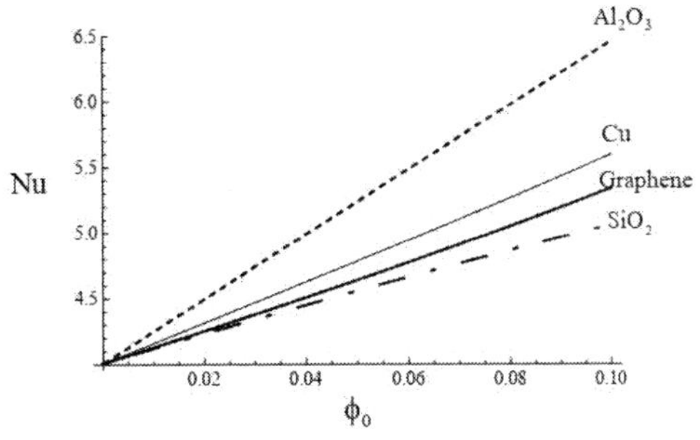

Figure 6. Nu as a function of volume fraction for the fourth kind of boundary condition (Gr/Re=1).

Conclusion

Analytical solutions for mixed convection heat transfer of a nanofluid between two parallel plates in the fully developed region has been presented for first and fourth kinds of boundary conditions. For the first kind, the pressure gradient was found to be independent of nanoparticle thermal conductivity. Results also show that SiO_2 gives the minimum pressure drop inside the channel while Al_2O_3 gives the maximum heat transfer rate.

References

[1] Hwang K S, Jang S P and Choi S U S 2009 Flow and convective heat transfer characteristics of water-based Al2O3 nanofluids in fully developed laminar flow regime *Int. J. Heat Mass Transf.* **52** 193–9
[2] Buongiorno J 2005 Convective transport in nanofluids *J. Heat Transf.* **128** 240–50 DOI: 10.1115/1.215083
[3] Grosan T and Pop I 2012 Fully developed mixed convection in a vertical channel filled by a nanofluid *J. Heat Transf.* **134** 082501
[4] di Schio E R, Celli M and Barletta A 2013 Effects of Brownian diffusion and thermophoresis on the laminar forced convection of a nanofluid in a channel *J. Heat Transf.* **136** 022401
[5] Yang C, Li W, Sano Y, Mochizuki M and Nakayama A 2013 On the anomalous convective heat transfer enhancement in nanofluids: A theoretical answer to the nanofluids controversy *J. Heat Transf.* **135** 054504
[6] Li W and Nakayama A 2015 Temperature dependency of thermophysical properties in convective heat transfer enhancement in nanofluids *J. Thermophys. Heat Transf.* **29** 504–12

Nomenclature

b	channel spacing
d_p	nanoparticle diameter
D_B	Brownian diffusivity, $= K_{BO}T/3\pi\mu_{bf}d_p$
D_T	thermophoresis diffusivity, $= 0.26k_{bf}/(2k_{bf}+k_p)*\mu_{bf}/\rho_{bf}*\phi_0$
Gr	Grashof number, $\frac{g\beta_{bf}q_1b^4}{\upsilon^2K_{bf}}$
K	thermal conductivity
K_{BO}	Boltzmann constant
Kr	solid/fluid thermal conductivity ratio, K_p/K_{bf}
N_{BT}	ratio of Brownian and thermophoretic diffusivities, $=\frac{D_B}{D_T}*\frac{T_0\phi_0}{\Delta T}$ for first kind, or $\frac{D_B}{D_T}*\frac{T_0\phi_0k_{bf}}{q_1b}$ for fourth kind
Nu	Nusselt number
p	fluid pressure at any cross section
p'	pressure defect at any cross section, $p-p_s$
p_0	fluid pressure at channel entrance
p_s	hydrostatic pressure, $-\rho_0 gz$
P	dimensionless pressure at any cross section, $\frac{p'-p_0}{\rho_0u_0^2}$

IOP Conf. Series: Journal of Physics: Conf. Series **829** (2017) 012018 doi:10.1088/1742-6596/829/1/012018

Re Reynolds number, $= \frac{u_o b}{\upsilon}$
T temperature at any point
T_0 inlet temperature
u_o entrance axial velocity
u longitudinal velocity component at any point
U dimensionless longitudinal velocity, $= u/u_o$
y horizontal coordinate
Y dimensionless horizontal coordinate, y/b
z vertical coordinate
Z dimensionless vertical coordinate, z/(b Re)

Greek Symbols
υ kinematic fluid viscosity
ρ fluid density
μ dynamic fluid viscosity
θ dimensionless temperature at any point, $= \frac{T - T_0}{T_1 - T_0}$ for first kind, or $\left[\frac{k_{bf}(T - T_m)}{q_1 b}\right]$ for fourth kind
β thermal expansion coefficient
ϕ particle volume fraction
Φ rescaled nanoparticle volume fraction, $\left[= \frac{\phi}{\phi_0}\right]$
γ immersed-particle buoyancy parameter, $= 0.78\pi \frac{\mu_{bf}^2}{\rho_{bf}} \frac{d_p}{K_{BO}\beta_{bf}T_0^2}(\frac{\rho_p}{\rho_{bf}} - 1)$

Subscripts

bf base fluid
nf nanofluid
w wall
p particle
1 duct wall at Y = 0
2 duct wall at Y = 1
0 condition at the entrance

IOP Conf. Series: Journal of Physics: Conf. Series **829** (2017) 012019 doi:10.1088/1742-6596/829/1/012019

Impedance spectroscopy of heterojunction solar cell a-SiC/c-Si with ITO antireflection film investigated at different temperatures

V Šály[1], M Perný[1], F Janíček[1], J Huran[2], M Mikolášek[1] and J Packa[1]

[1] Faculty of Electrical Engineering and Information Technology, Slovak University of Technology, Ilkovičova 3, 81219 Bratislava, Slovakia, Jozef.Huran@savba.sk
[2] Institute of Electrical Engineering, Slovak Academy of Sciences, Dúbravská cesta 9, 84104 Bratislava, Slovakia, vladimir.saly@stuba.sk

Abstract. Progressive smart photovoltaic technologies including heterostructures a-SiC/c-Si with ITO antireflection film are one of the prospective replacements of conventional photovoltaic silicon technology. Our paper is focused on the investigation of heterostructures a-SiC/c-Si provided with a layer of ITO (indium oxide/tin oxide 90/10 wt.%) which acts as a passivating and antireflection coating. Prepared photovoltaic cell structure was investigated at various temperatures and the influence of temperature on its operation was searched. The investigation of the dynamic properties of heterojunction PV cells was carried out using impedance spectroscopy. The equivalent AC circuit which approximates the measured impedance data was proposed. Assessment of the influence of the temperature on the operation of prepared heterostructure was carried out by analysis of the temperature dependence of AC equivalent circuit elements.

1. Introduction

Increasing interest regarding SiC is coming from space and military industries at the half of last century. Except of hard industry, high power and high temperature electronics silicon carbide has found application in photovoltaics, as well. SiC is very hard, so mechanically resistant. It crystallizes in more forms which offer variety of material properties and possible applications. Basically, it is IV-VI wide band-gap (over 2 eV) semiconductor [1]. Physical properties like thermal conductivity, high breakdown voltage, optical properties and mechanical resistance make SiC a suitable material applicable in extreme conditions and environment.

Crystalline layers of SiC growth at temperatures a lot higher than 1000 °C but the amorphous silicon carbide (a-SiC) growths at temperatures technically and energetically appropriate low – few hundreds degrees. In photovoltaics hydrogenated amorphous silicon carbide is known for its use as the window layer of high efficiency crystalline silicon solar cells creating heterojunction but the high light absorption in the range 400 to 600 nm leads to the limitations in short circuit current density. Improvements were reported when applying e.g. nanocrystalline SiC [2]. Amorphous silicon carbide could be also relevant substitute for amorphous silicon (a-Si:H) in silicon heterojunction solar cells (SHJ).

Transparent conductive oxides used at production of electronic devices are mostly required to have high electrical conductivity and optical trasmittance. The structural uniformity and deposition process which is compatible with substrate is important. Among conductive oxides indium-tin oxide (ITO) is widely used in PV technology. Microstructure, optical and electrical properties of ITO thin film, important when used for heterojunction solar cells, can be governed by the nature of the substrate, production conditions and post deposition treatment [3]. Moreover, the creation of parasitic Schottky barrier, also depended on deposition conditions, shall be avoided. ITO layer in PV structures is functioning mostly as antireflection coating or electrode. Adding ZnO to In_2O_3 results in an alloy commonly referred to as IZO.

Progressive smart photovoltaic technologies including heterostructures a-SiC/c-Si with ITO antireflection film are one of the prospective replacements of conventional photovoltaic silicon technology. The main advantages include acceptable efficiency, material savings, environmentally friendly production, acceptable service life and cost [4, 5, 6]. Similarly, as in the case of the conventional silicon technology, great emphasis must be placed on the appropriate arrangement of heterostructures into form of PV module and total reliability including used encapsulant material [7].

Our paper is focused on the investigation of heterostructures a-SiC/c-Si provided with a layer of ITO (indium oxide/tin oxide 90/10 wt.%) which acts as a passivation and antireflection coating.

2. Experimental

Thin a-SiC:H layer was prepared by plasma CVD deposition. A crystalline silicon wafer of (100) orientation was used as the substrate for the growth. The deposition was from the gas mixture of SiH_4, CH_4, H_2 and Ar, common for all prepared samples. The preparation process was optimized to improve the PV performance of final structures and is described in more details which one can find in [8]. The relationship between the fabrication parameters, particularly the change of flow rates SiH_4/CH_4, and the electric properties of a-SiC/c-Si samples prepared by PECVD was shown to be important and optimized. Proper concentration of CH_4 allows to obtain so called silicon rich amorphous carbide, which has good physical properties and was used as an emitter for SHJ. Antireflection coating formed by transparent conductive oxide (TCO) with defined thickness was suggested to improve PV response. ITO (indium oxide/tin oxide 90/10 wt.%) film was RF magnetron sputtered on top of the different phosphorus doped a-SiC:H(n) films. The thickness of the deposited ITO films was kept at 100 nm. Fig. 1 presents the final structure. All ITO films exhibits transmittance over 80 % in the region of 450 - 1000 nm. Comparing samples with different sputtering power P_{sput} (sputtering current I_{sput}) the highest transmittance is observed for sample prepared at P_{sput} = 100 W (Fig. 2) [9].

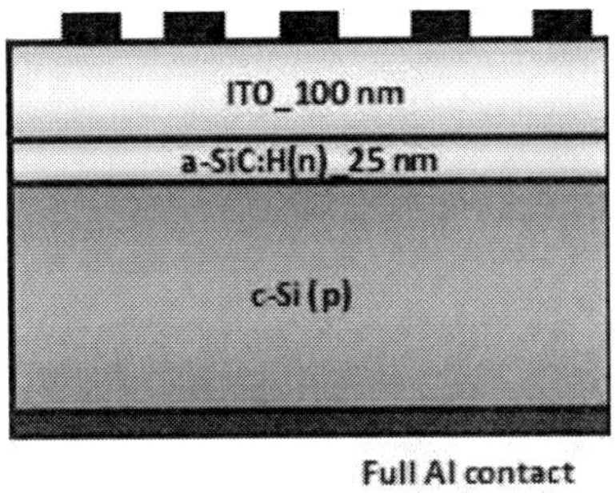

Figure 1. Schematic structure of SHJ sample.

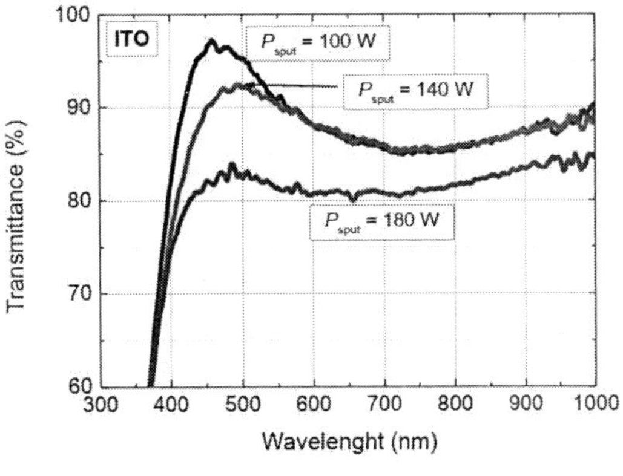

Figure 2. Optical transmittance of ITO layers prepared with varied sputtering power P_{sput}.

Photolitography and lift off technique was used for Al grid preparation as top contact of solar cell structure. Bottom Al ohmic contact was prepared by Al evaporation. Impedance spectra were measured by the Agilent LCR Meter 4284A. Samples were placed on a heated sample holder for purposes of AC measurement at various temperatures.
The equivalent AC circuit which best approximates the measured impedance data as a result of numerical process was proposed. Assessment of the effect of the temperature on the operation of prepared heterostructures was carried out by analysis of the temperature dependence of AC circuit elements. For a more comprehensive characterization the activation energies of conduction process were also determined. Impedance spectra were measured at different DC bias voltages due to a more detailed understanding of transport phenomenon in the heterostructure.

3. Results and discussion

The investigation of the dynamic properties of heterojunction PV cell was carried out. The Agilent LCR Meter 4284A was used to obtain data for impedance spectroscopy. The complex impedance at different temperatures was measured at small 25 mV AC signal. The obtained results are presented in the form of Nyquist plot. The influence of temperature is easy to quantify. Measured impedance spectra in the form of Nyquist plot in our case are represented by the superposition of two relaxation processes, but not simply distinguishable in Fig. 3, each for a different frequency interval. The presence of two semicircles, respectively two different relaxation processes, was recognized and confirmed by numerical simulations. Chosen impedance spectra – Nyquist characteristics – obtained on sample coated with ITO are shown in Fig. 3.

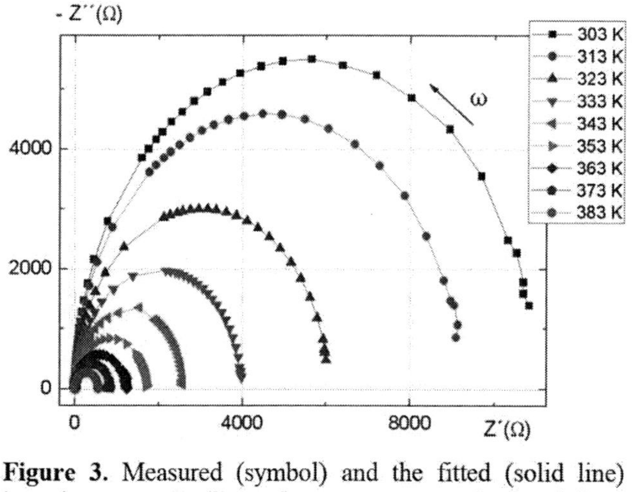

Figure 3. Measured (symbol) and the fitted (solid line) impedance spectra for various temperatures in the dark at DC bias 0 V.

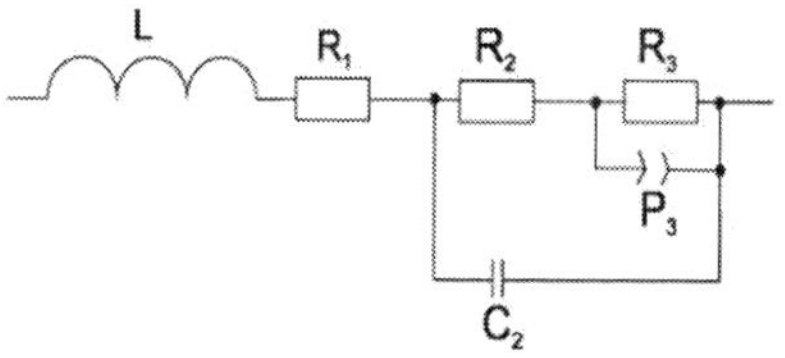

Figure 4. Proposed AC equivalent circuit.

The equivalent AC circuit which best approximated the measured impedance data, as a result of numerical procedure, was proposed. Assessment of the effect of the temperature on the behavior of prepared heterostructure was carried out by analysis of the temperature dependence of AC circuit elements. The temperature has a significant effect on the value of the shunt resistance (diameter of semicircle) as one can see in Fig. 3. The increase of temperature to 383 K from room temperature is reflected in the decrease of shunt resistance from appr. 12 kΩ to 500 Ω.
Impedance data were approximated using the equivalent circuit shown in Fig. 4. Very good correlation between the proposed AC circuit and the measured impedance data over the frequency range from 100 Hz to 1 MHz and at investigated temperatures interval by numerical simulation was found.

IOP Conf. Series: Journal of Physics: Conf. Series **829** (2017) 012019 doi:10.1088/1742-6596/829/1/012019

Elements of the equivalent circuit have a direct relation to electrical transport processes in the structure. Inductance L in the AC equivalent circuit represents inductance of leads and top finger grid system and in our case does not play any more important role. Series resistance R_1 is related to resistance of contacts and electrode system (grid on top and whole bottom area). Combination of resistance R_2 (transport resistance) and capacitance C_2 is related to space charge region of the pn hetero-junction situated within the structure. The combination of resistance R_3 (recombination resistance) and constant phase element (CPE), at Fig. 4 denoted as P_3, should be associated with traps and defects in the structure [10, 11].
The parallel combination of R_2 and C_2 is associated with the left part of Nyquist plot, that one measured in the area of higher frequencies. Semicircle assigned to R_3 and P_3 in lower frequency region is associated with internal charge transfer including accumulation (P_3) and recombination (R_3).
Constant phase element P_3 in AC equivalent circuit has been ascribed to and model different types of structural defects such as traps and structural inhomogeneities. Electrical impedance of CPE can be expressed as

$$Z_{CPE} = Y_0^{-1}(j\omega)^{-n}$$

where Y_0 represents CPE coefficient and n represents the character and an ideality of the distribution of relaxation times, ω is angular frequency [12, 13].
Detailed analysis of the CPE capacity exponent n on the applied DC bias enables the identification of transport processes in the structure. As can be seen in Fig.5, increased temperature do not resulted in any noticeable change of the exponent n for reverse and small forward biases. The presence of constant phase element instead of capacitance is confirming the frequency dispersion originating from disorder, porosity or inhomogeneity of materials. CPE exponent n ranges from 0 to 1. 1 is valid for pure capacitance. Decrease of the value of n at positive bias 200 mV accompanied by an increase of P_3 we ascribe to the distributed character of minority carriers movement and narrowing the space charge regionat higher positive bias voltage.

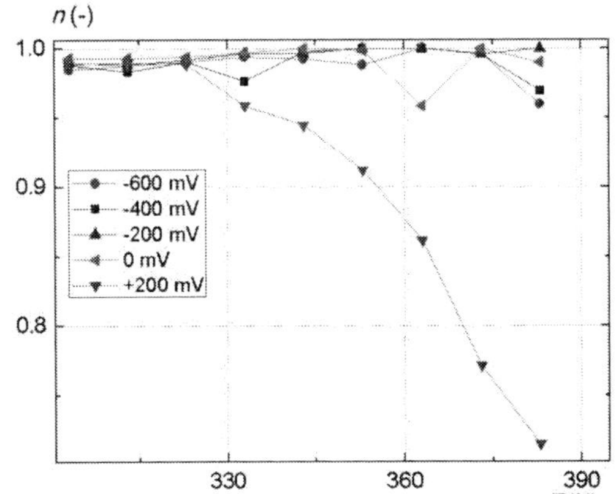

Figure 5. CPE impedance exponent n as a function of applied temperature.

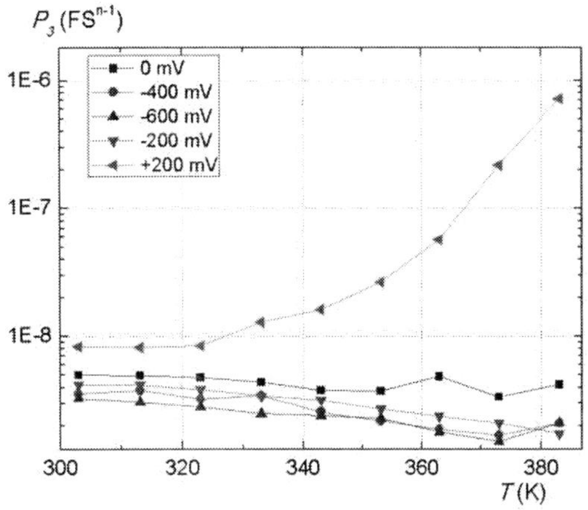

Figure 6. CPE capacitance as a function of applied temperature.

For further clarification the temperature depended impedance response the constant phase element at different bias voltage is shown in Fig.6. Reverse and small forward biases do not influence CPE capacitance value in any more remarkable way.

IOP Conf. Series: Journal of Physics: Conf. Series **829** (2017) 012019 doi:10.1088/1742-6596/829/1/012019

At higher forward voltage, that is at voltages closer to open circuit value, the depletion region associated with space charge region is disappearing (the capacitance C_2 is not more dominant) therefore, the equivalent circuit evolves to a circuit of parallel R_3 and CPE in series with series resistance R_1.
The dependence response of C_2 capacitance at different temperatures and applied bias is shown in Fig. 7.

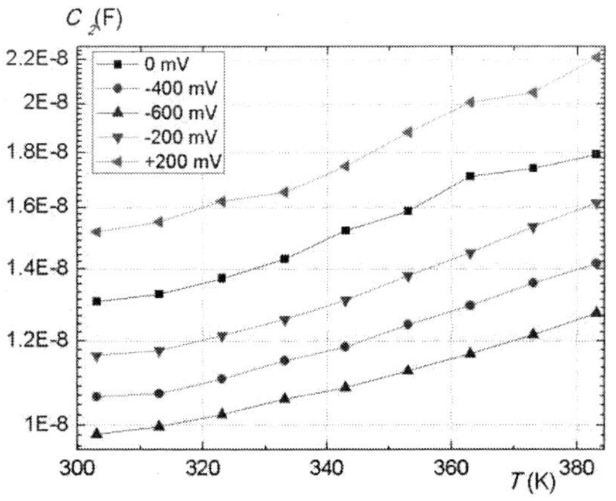

Figure 7. Parallel capacitance C_2 as a function of applied temperature.

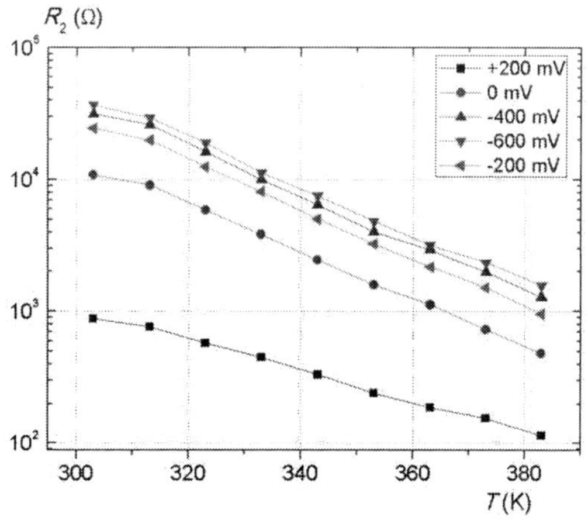

Figure 8. Parallel resistance R_2 as a function of applied temperature.

Fig. 8 shows the dependence of obtained parallel resistance R_2 on different applied temperatures at various bias voltages. In accordance with other studies [5, 6] we suppose that the resistance R_2 which is strongly dependent on applied temperature (Fig. 8), is transport resistance. The influence of the temperature regarding recombination resistance R_3 was minimal.
The increasing temperature activates the conductivity $G_2 = 1/R_2$. For a more comprehensive characterization the activation energies related to R_2 were calculated. The results are summarized in Tab. I.

Table I. Calculated values of activation energies for different DC bias.

DC bias (mV)	Ea (eV)
- 600	0.410
-400	0.413
- 200	0.417
0	0.402
200	0.263

4. Conclusion

The amorphous silicon carbide layer in heterostucture with crystalline silicon was prepared by PECVD method was optimized ratio of CH_4 and SiH_4. Antireflection coating formed by transparent conductive oxide with defined thickness was RF magnetron sputtered to improve PV response. The complex impedance measured in the dark using small AC signal was used to show the impact of temperature on the impedance spectra and electrical transport properties and characterize a-SiC/Si heterostructures. Equivalent circuit was proposed. Analysis of the temperature dependencies of its individual elements allows deeper insight into the behaviour of studied solar cell structure and suggest prospective improvements in technology steps and material properties.

Acknowledgments
This work has been supported by the Slovak Research and Development Agency under the contract No. APVV-15-0326. This work was also supported by the Scientific Grant Agency of the Ministry of Education of the Slovak Republic and of the Slovak Academy of Sciences under Project VEGA1/0651/16.

References

[1] Camassel J, Contreras S and Robert J 2000 *C. R. Acad. Sci. Paris*, t.1, **Serie IV**, Solids, fluids: electronic and optical properties 5-21
[2] Hamashita D, Kurokawa Y and Konagai M, *Energy Procedia* 2001 **10** 14-15
[3] Lien S Y, Lee C C, Wuu D S, Horng R H and Yeh M Y 2010 *Thin Solid Films* Characterization and optimization of ITO thin films for application in heterojunction silicon solar cells., **518** 10-13
[4] Yunaz I A, Hashizume K, Miyajima K, Yamada A and Konagai M 2009 *Solar Energy Materials & Solar Cells* **93** 1056–1061
[5] Janz S, Reber S and Glunz W 2006 *Proceedings of the 21st EUPVSEC*, 660-663
[6] Dao V A, Choi H, Heo J, Park H, Yoon K, Lee Y, Kim Y, Lakshminarayan N and Yi J 2010 *Current Applied Physics* **10** 506–509
[7] Polanský R, Pinkerová M, Bařinková M and Prosr P 2013 *Journal of Electrical Engineering – Elektrotechnický časopis* Mechanical behaviour and thermal stability of EVA encapsulant material used in photovoltaic modules **64** 361 – 365
[8] Šály V, Huran J, Mikolášek M, Perný M and Váry M 2014 *Proc. 29th PVSEC* AC and DC characterization of Al/a-SiC/c-Si(p)/Al structures under dark for PV applications. In:. Amsterdam Netherlands 84
[9] Mikolášek M, Harmatha L, Perný M, Šály V and Huran J 2016 *Proc. Power engineering 2016 Renewable Energy Sources 2016* Development of silicon heterojunction solar cells with IZO and ITO antireflection coatings 6th International Scientific Conference Slovakia May 31 - June 2 104-109
[10] Leever B J, Bailey C A, Marks T J, Hersam M C, and Durstock M F 2012 *Adv. Energy Mater.* In situcharacterization of lifetime and morphology in operating bulk heterojunction organic photovoltaic devices by impedance spectroscopy **2** 120-128
[11] Bisquert and Garcia-Belmonte G 2011 *J. Phys. Chem. Lett.* On voltage, photovoltage, and photocurrent in Bulk Heterojunction Organic Solar Cells **2** 1950-1964
[12] Barsoukov E and Macdonald J R 2005 *Impedance Spectroscopy Impedance Spectroscopy: Theory, Experiment, and Applications* 2nd Edition John Wiley & Sons
[13] Hsu C H and Mansfeld F 2001 *Corrosion* Concerning the conversion of the constant phase element parameter Y0 into a capacitance **57** 747

Some Features of Pressure Evolution in Systems "Non-Wetting Liquid – Nanoporous Medium" at Impact Intrusion

V A Byrkin, A A Belogorlov, D A Paryohin and A S Mitrofanova

National Research Nuclear University MEPhI (Moscow Engineering Physics Institute), Kashirskoe highway 31, Moscow, 115409, Russia

E-mail: VAByrkin@mephi.ru

Abstract. The last few decades systems consisting of nanoporous medium dispersed in a non-wetting liquid cause an increased interest from both the practical and theoretical points of view. Non-wetting liquid can infiltrate into the porous medium only with an excess pressure. Liquid infiltration tends to increase the solid-liquid interfacial energy and the absorbed energy is proportional to the specific surface area of the medium. Therefore this energy for nanoporous media can reach several orders of magnitude superior to traditional damping materials and shape-memory materials. As a consequence, the prospects of using devices based on systems consisting of a nanoporous medium immersed in a non-wetting liquid associated mainly with the absorption of mechanical energy of impact or explosion. The paper presents the results of experimental studies of impact intrusion the systems of industrially produced hydrophobic silicas Fluka 100 C8 and Fluka 100 C18 with distilled water. With increasing the impact energies nontrivial pattern of pressure changes in the system over time is observed.

1. Introduction

For dispersing the non-wetting liquid into the system of liquid clusters in the pores of the nanoporous disordered medium external pressure must be applied. Depending on the rate of growth exrernal pressure [1] two different scenarios are possible to fill the pores with liquid. Wherein physics infiltration process at high speeds the pressure is significantly different from infiltration at low speeds. Most modern experimental and theretical study of such systems is related to the quasi-static regime with the infiltration rate of the pressure of not more than $\dot{p} < 10$ atm/s. At such speed it was found a number of effects, such as a dispersion transition [2] and anomalously slow relaxation [3]. These effects open up a range of opportunities for different applications: sensors, passive protection systems with controlled release of substances such as drugs with changes in temperature, the passive fire protection systems in creating smart materials with controllable permeability for liquids or gases.

At impact or explosion a rapid growth of pressure is occured in system with rates of $\dot{p} = 10^4 \div 10^5$ atm/s. The prospects of using devices based on systems consisting of a nanoporous medium immersed in a non-wetting liquid associated mainly with the absorption of mechanical energy of impact or explosion. Liquid infiltration tends to increase the solid-liquid interfacial energy and thus the absorbed energy can be estimated as $E \sim \Delta\gamma S$, where $\Delta\gamma$ is a solid-liquid interfacial tension, and S is the specific surface area of the medium. Therefore this energy for nanoporous media can reach several orders of magnitude superior to traditional damping materials and shape-memory materials [4, 5]. Particularly there are considered the development of automotive dampers [6], the system of protection against strains and by the earthquake [7]. To the theoretical study of the processes by rapid intrusion, including modeling, devoted not so much articles [1, 8–10]. In [8] shows that the observable system can absorb more energy density in comparison with the case of quasi-static pressure. A lot of papers [6] are devoted to experimentally investigation of the development of car damper. Despite this, a set of experimental data is presented rather poorly and full-fledged studies are missing.

2. Experiment

In the present study we investigated the dynamics of non-wetting liquid during water infiltration into commercially available silica gels with a random structure of pores obtained in the sol-gel process Fluka 100 C8 — Silica gel 100 C8, Reversed phase (Sigma-Aldrich catalogue 60755) and Fluka 100 C18 — Silica gel 100 C18, Reversed phase (Sigma-Aldrich catalogue 60756-50G). Using the pycnometry (micro-Ultrapyc 1200e, Quantachrome Instruments) and porometry (Nova 1200e, Quantachrome Instruments) methods, we determined the density (ρ), specific surface area (S_p), specific volume of pores (V_p), and pore size distribution by the classical Barrett-Joyner-Halenda (BJH) method within the cylindrical model of pores and porosity (φ) of this material. The results are presented in Table 1.

A 4 g sample of the nanoporous medium was placed in a liquid-permeable container in the pressure chamber with volume of $\sim$ 60 cm^3 (Fig. 1). The chamber was filled with distilled water. A falling weight from certain height is designed to impart a load pulse. Impact energy is determined by the variation of height. The direct method of measurement is described in detail at [1].

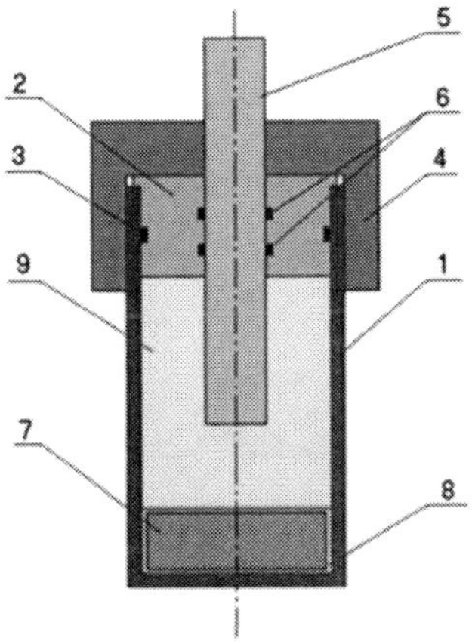

Figure 1. The pressure chamber: 1 cell; 2 plug; 3, 6 gaskets; 4 screw top; 5 stock; 7 nanoporous medium; 8 membrane; 9 non-wetting liquid (water)

The typical results of experimental measurements of pressure versus time with nanoporous media are shown in Fig. 2, 3, the range of variation of energy ranged from 10 to 60 J.

IOP Conf. Series: Journal of Physics: Conf. Series **829** (2017) 012020 doi:10.1088/1742-6596/829/1/012020

Table 1. Characteristics of the studied porous media.

	Fluka 100 C8	Fluka 100 C18
$\bar{R}$, nm	3.9 ± 0.2	3.9 ± 0.2
ρ, g/cm^3	1.7603 ± 0.0034	1.6125 ± 0.0025
V_p, cm^3/g	0.56 ± 0.02	0.46 ± 0.02
S_p, m^2/g (BET)	267 ± 10	183 ± 6
$\varphi = V_p/(V_p + 1/\rho)$	0.49 ± 0.02	0.42 ± 0.02

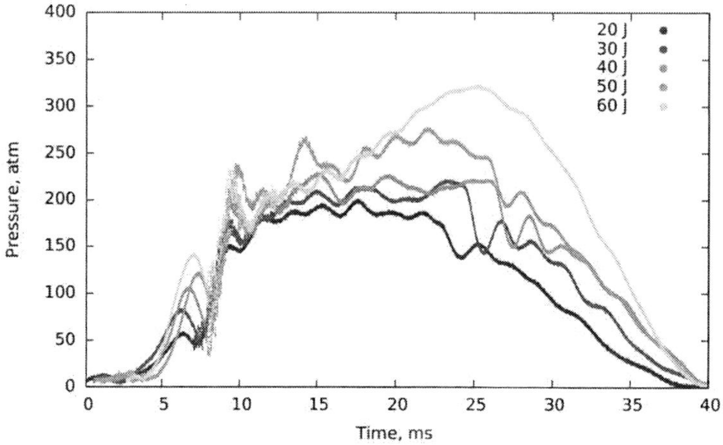

Figure 2. The dependence on the time of pressure change in the system "nanoporous medium – non-wetting liquid" for Fluka 100 C8 – water at different impact energies

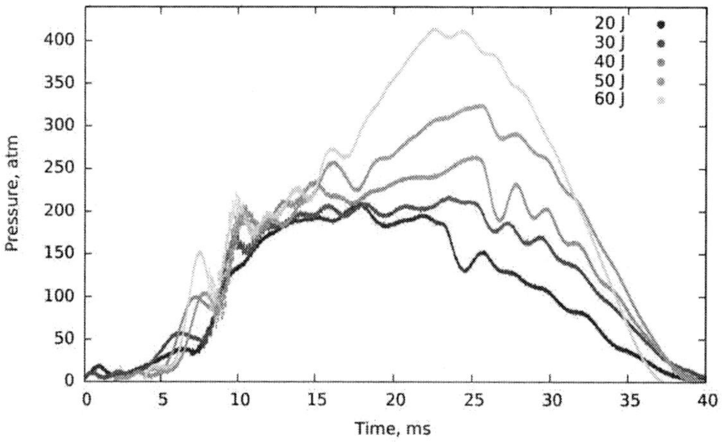

Figure 3. The dependence on the time of pressure change in the system "nanoporous medium – non-wetting liquid" for Fluka 100 C18 – water at different impact energies

3. Results and disscussion

Evolution of pressure occurs in several stages and depends essentially on the energy of impact. If the energy is not enough to overcome the pressure threshold at impact intrusion p_0, then there is an elastic compression of the entire system without filling pores of a porous medium. For the systems "granular porous medium Silohrom CX 1.5 – Wood's alloy", "Libersorb 23(C8) – water" and "Fluka 100 – water", it was found that infiltraion occurs at the percolation threshold at a pressure far above the threshold pressure at quasi-static mode p_{c0} [1]. So far the first system pressure threshold value of $p_0 = 1.6p_{c0}$ [1], and for the third — $p_0 = 2p_{c0}$ [8]. If the energy is such that the pressure exceeds the value of p_0, then before system achieve this pressure also elastic compression occurs, which is then followed by the filling of pores by non-wetting liquid. In this case intrusion is occurred at constant pressure. These experiments (Fig. 2, 3) shows that at the energy of 10 J pressure is enough to the beginning of filling at impact intrusion. At higher energies, where the energy of compression exceeds the maximum value determined by the energy density of the porous medium filling and intrusion mode at constant pressure is not realized. In these circumstances a response of system non-wetting liquids – nanoporous medium is elastic deformation due to the filling of the porous medium is not provided the necessary speed of energy absorption and characteristic pressure growth time. This case is accompanied at the end of filling increase of pressure to a maximum and then only outflow.

A intrusion pressure dependence on the impact energy of the investigated porous media is presented in Fig. 4. Data are also presented for the maximum pressure. From these data it follows that the intrusion pressure changes slowly with a slight tendency to increase. Lastly, maximal pressure for Fluka 100 C18 grows two times.

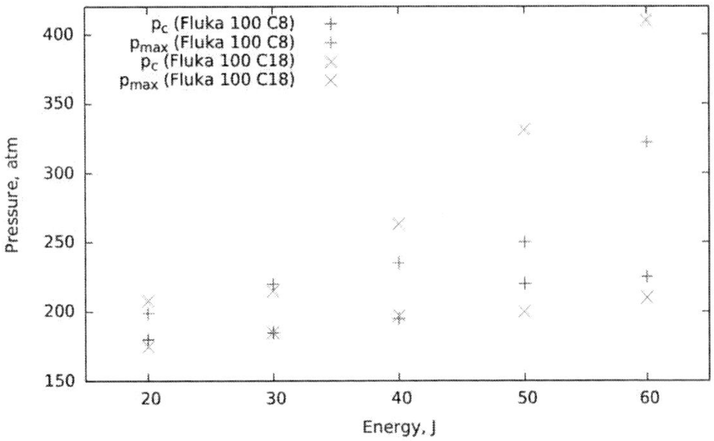

Figure 4. The intrusion pressure and maximum pressure dependences on energy in the system "nanoporous medium – non-wetting liquid"

Another feature of pressure evolution is the presence of oscillations. Such oscillations have been observed previously for the systems "Silohrom CX 1.5 – Wood's alloy" and "Libersorb 23 (C8) – water" [1, 12] in the range of energy, where a constant pressure is occurred. However, the reason why they are not observed in the studied systems in this paper remains open and requires a separate study.

Thus, the magnitude of the change in volume of the system within the measurement error coincides with a change in volume due to deformation, and this means that there is no filling of the porous medium at low impact energies. With increasing impact energy is observed infiltration to the porous medium with a pronounced plateau in the pressure versus time. At higher energies intrusion mode at constant pressure is not realized due to the filling of the porous medium is not provided the necessary speed energy absorption. The response of system non-wetting liquids – nanoporous medium in this case is elastic deformation.

Acknowledgments
The reported study was partially supported by RFBR.

References
[1] Borman V D, Belogorlov A A, Lisichkin G V, Tronin V N and Troyan V I 2009 *J. Exp. Theor. Phys.* **108** 389–410
[2] Borman V D, Belogorlov A A, Byrkin V A, Tronin V N and Troyan V I 2012 *JETP Lett.* **95** 511–14
[3] Borman V D, Belogorlov A A and Tronin V N 2016 *Phys. Rev. E* **93** 022142
[4] Chang K C, Soong T T, Lai M L and Nielsen E J 1993 *Earthquake Spectra***9** 371–87
[5] Eroshenko V, Grosu Y, Tsyrin N, Stoudenets V, Nedelec J M and Grolier J P E 2015 *J. Phys. Chem. C* **119** 10266–72
[6] Suciu C V and Buma S 2013 *Proc. of the FISITA 2012 World Automotive Congress* (*Lect. Notes Electr. Eng.* **198**) (Berlin: Springer) pp 21–32
[7] Chen X, Surani F B, Kong X, Punyamurtula V K and Qiao Y 2006 *Appl. Phys. Lett.* **89** 241918
[8] Surani F B, Kong X, Panchal D B and Qiao Y 2005 *Appl. Phys. Lett.* **87** 163111
[9] Liu H and Cao G 2013 *J. Chem. Phys.* **139** 114701
[10] Liu H and Cao G 2014 *J. Phys. Chem. C* **118** 25223–33
[11] Belogorlov A A, Byrkin V A, Borman V D and Tronin V N 2016 *J. Phys.: Conf. Ser.* **751** 012035
[12] Borman V D, Belogorlov A A, Grekhov A M et al.2001 *Jetp Lett.* **74** 258

IOP Conf. Series: Journal of Physics: Conf. Series **829** (2017) 012021 doi:10.1088/1742-6596/829/1/012021

Magnetoresistance Aharonov–Bohm (MAB) oscillations in elongated quantum dots

K M Gambaryan*, V M Aroutiounian, V G Harutyunyan and L S Yeranyan

Department of Physics of Semiconductors and Microelectronics, Yerevan State University, 1 Alex Manoukyan, Yerevan 0025, Armenia

*E-mail: kgambaryan@ysu.am

Abstract. The type-II InAsSbP ellipsoidal quantum dots (QDs) are nucleated on InAs substrate from In-As-Sb-P quaternary melt-solution using Stranski–Krastanow growth mode. The structures under consideration were prepared in the form of photoconductive cells. Magnetospectroscopy is used to measure the QDs structure's electric sheet resistance in magnetic field at lateral current flow. The magnetoresistance Aharonov–Bohm (MAB) oscillations with the period of $\delta B \approx 0.4$ T are found. No temperature dependence is experimentally revealed for the period of MAB. It is shown the QDs size distribution strongly acts on the period of MAB oscillations. The values for both major and minor semiaxes of ellipsoidal QDs were theoretically calculated using an equation for the period of classical Aharonov-Bohm oscillations. Comparison of calculated and experimentally measured values shows that they are coincide with high accuracy. Additionally, the temperature dependant hysteresis on magnetoresistance curves is revealed.

1. Introduction

Three-dimensional confinement of carriers in a quantum dot (QD) transforms the continuous optical spectrum of a bulk semiconductor into a size-tunable, atomic-like spectrum featuring a series of sharp peaks associated with discrete electronic transitions. These unique electronic and optical properties of QDs have been exploited for a wide range of applications, including quantum electronics and quantum information processing [1, 2], optical communications, display technologies, solar energy harvesting and fluorescent labeling etc. [3]. No classical states of light also can be generated using semiconductor QDs [3-5], enabling on-chip semiconductor quantum optics. Among quantum size objects' fabrication techniques, when the difference between lattice constants of the growing materials is 3-10%, the self-organized Stranski–Krastanow (S–K) method can be realized, by which dislocation-free nanostructures can be produced [7]. In a strained film, the formation of islands is energetically favorable, as it reduces the strain energy in the crystal [8]. The fabrication and characterization of epitaxially-grown semiconductor QDs had advanced to the point that QD-based devices were realizable. Despite the rapid progress in this field, numerous questions remain with regard to how QD morphology affects the electronic and optical properties and coherent light-matter interactions in semiconductor QDs. The performance of QD-based devices relying on these interactions is typically limited by multiple effects that are still not understood: 1) short coherence and relaxation times due to interactions of the charge carriers with each other and the local environment; 2) sensitivity of the QD properties on size and confinement; and 3) the presence of many-body interactions that often dominate the linear and nonlinear optical response [3].

In classical mechanics, the electromagnetic potentials (φ and $\vec{A}$) first were introduced as a pure mathematical notions for calculating the fields. And the fundamental equations of motion can be expressed only in terms of fields. But in quantum mechanics the equations of motion of a particle are replaced by the Schrodinger equation for a wave, which cannot be expressed in terms of the fields alone, but also requires the electromagnetic potentials [9].

In this paper, actually we are experimentally revealed the magnetoresistance oscillations in InAsSbP type-II QDs due to Aharonov-Bohm effect. Additionally, theoretical calculations of QDs lateral dimensions using the measured period of MAB oscillations are performed.

2. Basic theory

The A-B phase is most simply described as the phase that appears in charged particle's wave function during its passage through the region where magnetic flux exists, but no effects of classical Lorentz force is present. In semiconductor nanostructures, such as type-II quantum dots or quantum rings, applied magnetic field penetrating the area surrounded by their trajectories will cause a phase shift of the constituent particle [10]. At zero magnetic field exciton ground state has zero angular momentum projection (L=0), but with increasing magnetic flux its changes to states with higher angular momentum ($L = -1,-2,-3$), causing oscillation of ground state energy of the charge carrier (figure 1*c*) [11].

Assuming that electron and hole move with different radii and the interaction between them is negligible, in the presence of external magnetic field, the energy of the electron-hole pair will be [9]:

$$E_{exc} = E_g + \frac{\hbar^2}{2m_eR_e^2}\left(l_e + \frac{\Phi_e}{\Phi_0}\right)^2 + \frac{\hbar^2}{2m_hR_h^2}\left(l_h - \frac{\Phi_h}{\Phi_0}\right)^2, \tag{1}$$

where E_g includes the band gap and confinement energies and is the magnetic field-independent term, $R_{e(h)}$ is the rotation radius of the electron (hole), l_e and l_h are the angular momentum (magnetic quantum numbers) of the electron and hole, respectively, Φ_0 is the flux quantum: $\Phi_0 = h/e$ and Φ_e and Φ_h magnetic fluxes are equal to $\Phi_e = \pi R_e^2 B$ and $\Phi_h = \pi R_h^2 B$, respectively. Now, taking in account strong Coulomb interaction, when they will rotate together as shown in Fig. 1(b), the exciton spectrum becomes [12]

$$E_{exc} = E_g + \frac{\hbar^2}{2MR_0^2}\left(L + \frac{\Delta\Phi}{\Phi_0}\right)^2, \tag{2}$$

where $L = l_e + l_h$ is the total momentum of an exciton, $R_0 = (R_h + R_e)/2$, and $M = (m_eR_e^2 + m_hR_h^2)/R_0^2$. Magnetic flux between the trajectories of the particles equals to:

$$\Delta\Phi = \Phi_e - \Phi_h = \pi B\left(R_e^2 - R_h^2\right). \tag{3}$$

In case of type-II QDs with strongly confined holes, we can assume that $R_h \to 0$, therefore $MR_0^2 = m_eR_e^2$ and $\Delta\Phi = \Phi_e$.

In direct-gap semiconductors emission process requires the total angular momentum to be equal 0. Since $l_h = 0$, all ground states of the electron in the conduction band with $l_e = -1, -2, -3...$ states are dark. Hence, the exciton emission from cylindrical type-II QDs in magnetic field will be blocked for $B > B_{AB}$, where B_{AB} corresponds to the field at which the electron ground state energy passes from $l_e = 0$ to $l_e = -1$. However, in real systems due to the presence of defects in QDs, above described mechanism for emission may not take place.

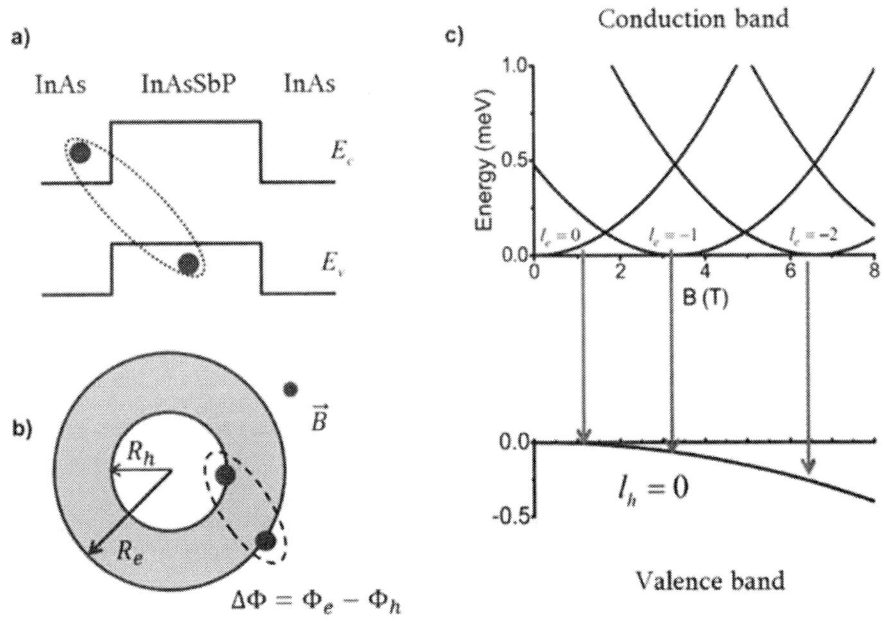

Figure 1. Schematic band structure of type-II QD (*a*). The sketch of an electron-hole pair in type-II quantum dot (*b*). The system energy spectra in magnetic field (*c*).

3. Experimental details

Investigated InAsSbP type-II QDs were grown on industrial undoped n-InAs(100) substrate by liquid phase epitaxy (LPE) using the Stranski–Krastanow growth mode at T=550°C constant temperature. The atomic force microscope (AFM «TM Microscopes–Autoprobe CP») and «Keithley-6514 System Electrometer» were used for morphological characterization and magnetospectroscopy, respectively. AFM images of ellipsoidal QDs which are under consideration are presented in figure 2.

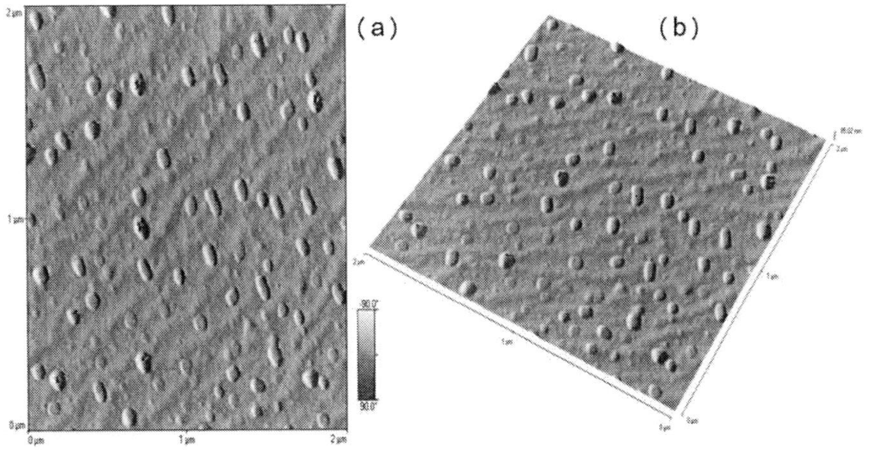

Figure 2. AFM images of InAsSbP type-II ellipsoidal QDs in plane (*a*) and oblique (*b*) view.

According to the InAs–InSb–InP phase diagram, concentrations of antimony and phosphorus in the growth solution were chosen to provide a lattice mismatch up to 3% between the InAs substrate and

InAsSbP wetting layer at T=550°C. Ohmic contacts of the samples were fabricated by the thermal vacuum evaporation technique. The sample's active surface was chosen to be equal to 10^{-2} cm^2. Measurements were performed at gradually increasing magnetic field up to 1.6 T with further decreasing up to zero, which was set to be perpendicular to the surface of quantum dots.

4. Results and discussions

Dependence of magnetoresistance on magnetic field measured at Faraday geometry at 300 K is presented in figure 3(*a*). The dependence of the derivative curve, on which observed effect is more noticeable, is presented in figure 3(*b*). Observed fractures in this curve, possibly are due to the transitions of charge carriers between energy levels due to the influence of magnetic field. As it was shown in [13], the magnetic flux, evidently, leads to a periodic change in the quantum mechanical properties of the encircling electrons. Therefore, under the influence of perpendicular magnetic field, the electron orbits the QD periphery producing an observable MAB, despite the selection rules for transitions in angular momentum, that are strictly valid only for the situation of perfect rotational symmetry and low temperatures.

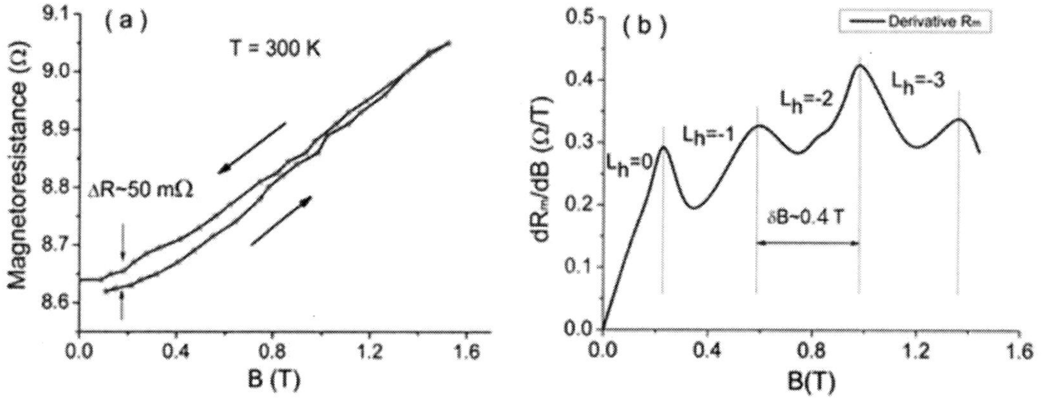

Figure 3. Type-II QDs structure's resistance versus magnetic field at room temperature (*a*) and its derivative (*b*).

From figure 3(*b*) also one can see, that the period of oscillations equals to ~0.38 T. Interestingly note that the results of measurements performed at liquid nitrogen temperature show the same value for the period of MAB. This result was expected, because the period of Aharonov-Bohm oscillations is given by following equation:

$$\delta B = \frac{\Delta\Phi}{\Phi_0} = \frac{h}{e\pi ab}, \tag{4}$$

where h is a Plank constant, e is an elementary charge, a and b are major and minor axles of ellipsoidal quantum dots, respectively, and there is no any temperature dependent parameters. Additionally, AFM measurements show that in our experiments a=1.23b. Using equation (4), dependence of the MAB period on ellipsoidal QDs axles was built, which 3D sketch is presented in figure 4.

Intersection points of a=1.23b line and projection curve of the cross section of 3D graph by $\delta B = const$ planes (figure 5(*a*)) show sizes of ellipsoidal quantum dots, in which the oscillations are take place at respective value of δB. In figure 5(*b*) we present intersection point of a=1.23b line and cross section line at $\delta B = 0.37T$ from which we can define average sizes of our quantum dots. On the other hand, for each ellipse we can find corresponding circle with the same surface area. In our case the radii of corresponding circle is equal to 59.7 nm (black curve) and for major and minor semiaxis 66 nm and 54 nm, respectively. At the same time, infinite number of ellipses correspondes to that line.

IOP Conf. Series: Journal of Physics: Conf. Series **829** (2017) 012021 doi:10.1088/1742-6596/829/1/012021

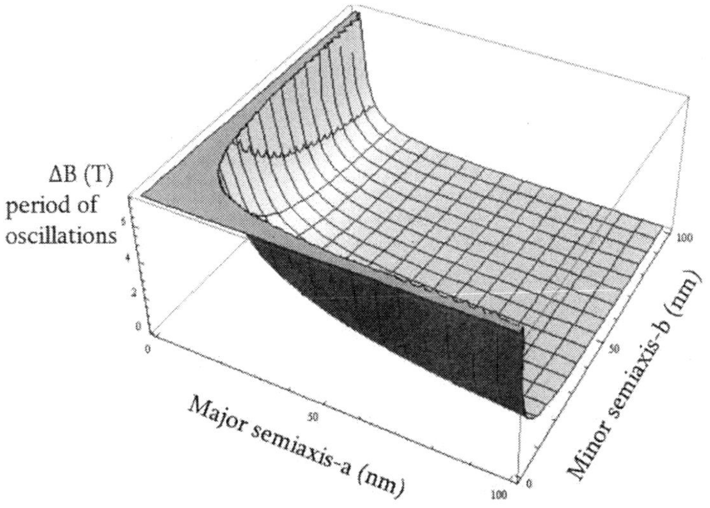

Figure 4. Dependence of MAB oscillations period on ellipsoidal QDs axles.

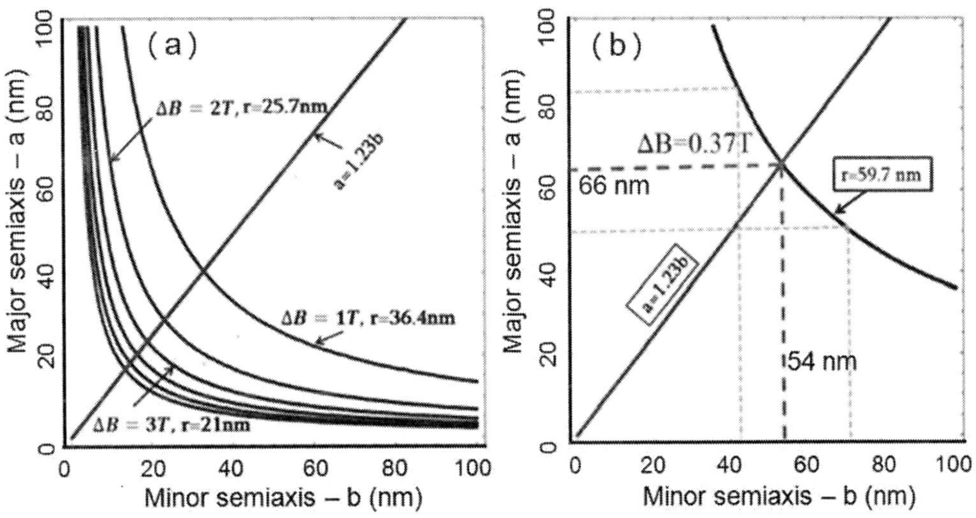

Figure 5. Cross sections of 3D graph by $\delta B = const$ – (a) and $\delta B = 0.37T$ – (b) planes.

Oblique view AFM images of single QDs grown from In-As-Sb-P quaternary liquid phase on InAs(100) substrate are presented in figure 6, from which one can see that the growth process and technological conditions result in an elongation of QDs in certain direction.

Comparing theoretically obtained values for quantum dots' sizes with experimentally measured values, it is not difficult to see, that they are coincide with high accuracy. This fact allows us to precisely define the size of cylindrical quantum dots in which MAB has occurred. As for ellipsoidal quantum dots, we only need to know the relation between values of semiaxes.

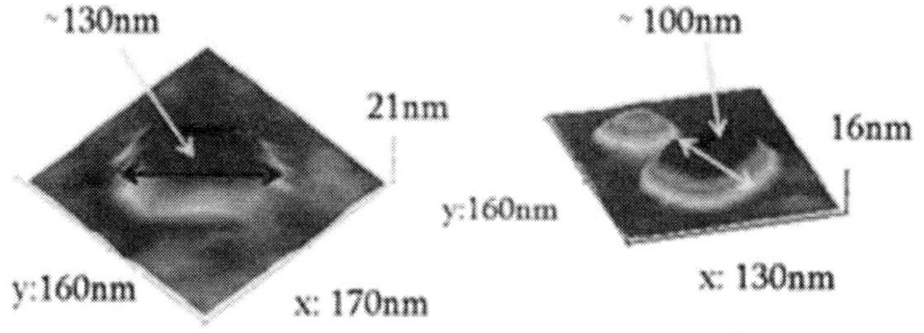

Figure 6. AFM images of QDs grown on InAs(100) substrate.

5. Conclusions

Thus, Aharonov-Bohm oscillations with the period of $\delta B \approx 0.4$ T were found on the curve of the magnetoresistance derivative at room temperature. The average values for both major and minor semiaxes of ellipsoidal QDs were theoretically calculated by the equation for the period of Aharonov-Bohm oscillations. Comparison of calculated and experimentally measured values shows that they are coincide with high accuracy.

Acknowledgments

This work was performed in the frame of the project № 15T-2J137 funded by the State Committee of Science of Armenia.

References

[1] Biolatti E, Iotti R, Zanardi P and Rossi F 2000 *Phys. Rev. Lett.* **85** 5647-5650
[2] Troiani F, Hohenester U and Molinari E 2000 *Phys. Rev. B* **62** 2263-2266
[3] Moody G 2013 *Confinement Effects on the Electronic and Optical Properties of Semiconductor Quantum Dots Revealed with Two-Dimensional Coherent Spectroscopy* (University of Colorado, PhD Thesis)
[4] Michler P, Kiraz A, Becher C, Schoenfeld W V, Petroff P M, Zhang L, Hu E and Imamoglu A 2000 *Science* **290** 2282
[5] Benson O, Santori C, Pelton M and Yamamoto Y 2000 *Physical Review Letters* **84** 2513
[6] Santori C, Fattal D, Vuckovic J, Solomon G S and Yamamoto Y 2002 *Nature* **419** 594
[7] Stranski I and Krastanow L 1937 *Springer* **71** 351-364
[8] Baskaran A and Smereka P 2012 *J. Appl. Phys.* **111** 1-6
[9] Aharonov Y and Bohm D 1959 *The Physical Review* **115** 485-491
[10] Fomin V 2014 *Physics of Quantum Rings* (Berlin, Springer)
[11] Govorov A, Ulloa S, Karrai K and Warburton R 2002 *Phys. Rev. B* **66** 1-4
[12] Kalameitsev A, Govorov A and Kovalev V 1998 *JETP Letters* **68** 669-672
[13] Gambaryan K M, Harutyunyan V G, Aroutiounian V M, Ai Y, Ashalley E and Wang Z M 2015 *J. Phys. D: Appl. Phys.* **48** 275302

IOP Conf. Series: Journal of Physics: Conf. Series **829** (2017) 012022 doi:10.1088/1742-6596/829/1/012022

NMR investigation of domain wall dynamics and hyperfine field anisotropy in magnets by the magnetic video-pulse excitation method

Ts A Gavasheli[1], GI Mamniashvili[2] and T O Gegechkori[2]

[1]Iv. Javakhishvili Tbilisi State University, Georgia

[2]E. Andronikashvili Institute of Physics, Iv. Javakhishvili Tbilisi State University,Georgia

tsismari.gavasheli@tsu.ge

Abstract. Two-pulse nuclear spin echoes were studied experimentally depending on the time of application and pulse amplitudes of the DC magnetic field-magnetic video-pulses (MVP) as well as on the value of the external magnetic field. The measurements were performed with nanopowders and polycrystals of metallic cobalt, in lithium ferrite and half metal Co_2MnSi. Two types of dependences of these signals on time of application of MVP with respect to moments of application of exciting radio-frequency pulses were established, which were determined by the degree of anisotropy of local hyperfine fields. The mechanisms of influence of the pinning and mobility of domain walls on the revealed specific features of the signals under study are also discussed. It is shown that temporal spectra of the MVP effect on two-pulse echoes in multidomain magnets are determined by the parameters of domain walls and can be used for qualitative and quantitative characterization of the domain wall dynamics of magnets.

For characterization of microscopic magnetic properties of magnets the method of MVP excitation was used [1] by which the influence of MVP field on the nuclear spin echo in magnetic materials was studied. The investigation includes the determination of nuclear spin echo dependences in a sample on the amplitude H_d and duration τ_m of MVP applied in the interval between RF pulses or between the second RF pulse and echo signal (asymmetric action), as well as, during RF pulse action, Figure1a (symmetric action).

The decrease of echo signal intensity at switching-on of the MVP in intervals between RF pulses or between RF pulse and echo signal is related with the breakdown of phase coherence in a system of processing nuclear moments because of change in the local nuclear frequencies caused by hyperfine field (HFF) anisotropy under the action of MVP displacing domain walls.

The switching-on MVP symmetric in respect of the second RF pulse results in the fact that the first and the second RF pulses successfully excite nuclei changing their positions in domain walls defining on their resonance frequencies and the RF field gain factor (η) [1]. At sufficiently short RF pulses the resonance frequency change could be neglected but the change of η results in the decrease of echo intensity.

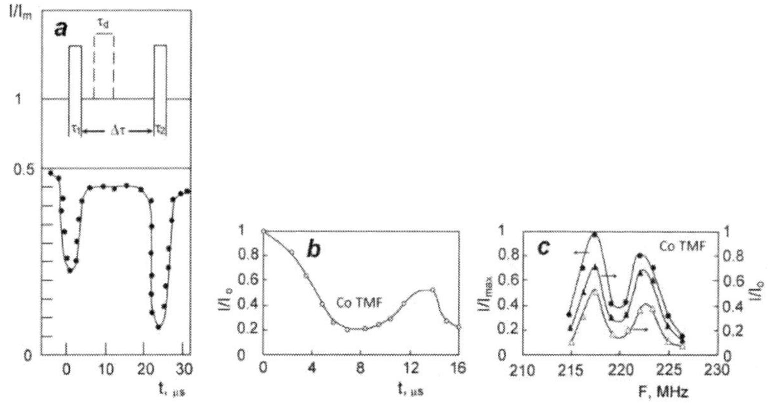

Figure 1. a) Timing diagrams of the relative intensity I/Imax dependence of the two-pulse echo (•) on the temporal location of a magnetic video-pulse of Hd=5 Oe, for 57Fe NMR in lithium ferrite at: τ1 = τ2 = 0.8 μs, Δτ = 21 μs, τd = 3 μs, fNMR=74.0 MHz, Hd=0. τ1, τ2, Δτ, τd are RF pulse durations, time interval between them and magnetic pulse duration, correspondingly;

b) Timing diagrams of the intensity dependence of the two-pulse echo on the temporal location of a magnetic video-pulse of Hd, in cobalt thin magnetic films at: τ1 = τ2 = 1.5 μs, Δτ = 9 μs, τd = 3 μs, Hd=10 Oe, fNMR=216.5 MHz, Io – echo amplitude at Hd=0;

c) NMR spectrum of Co film (♦) and frequency spectra diagrams for magnetic video-pulse influence for symmetric (▲) and asymmetric (□) application at: τ1 = 1.3 μs, τ2 = 1.5 μs, Δτ = 9 μs, τd = 3 μs, Hd=10 Oe, [1].

As it follows from Figures1–2, the essential difference takes place for the MVP action on the two-pulse echo in different magnets.

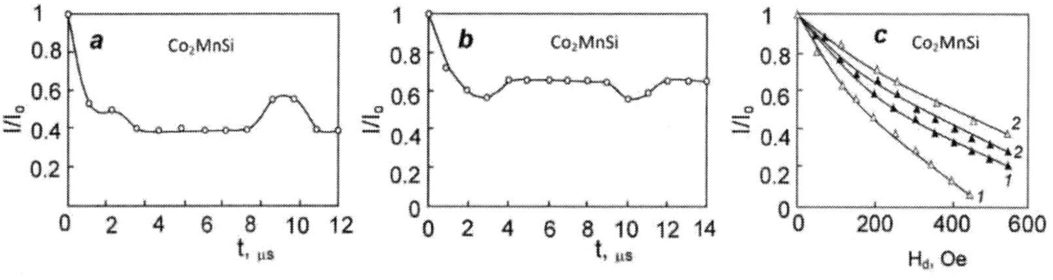

Figure 2. Timing diagrams (TD) of the intensity dependence of the two-pulse echo on the temporal location of the magnetic pulse with duration τ_d in Co_2MnSi (*a*) for ^{59}Co NMR at: τ_1 = 1.1 μs, τ_2 = 1.4 μs, $\Delta\tau$ = 10 μs, τ_d = 2 μs, f=145,5 MHz, H_d=550 Oe and (*b*) for ^{55}Mn NMR at: $\tau_1 = \tau_2$ = 3 μs, $\Delta\tau$ = 7 μs, τ_d = 2 μs, f=354 MHz, H_d=300 Oe.

c) Two-pulse echo intensity dependences on magnetic video-pulse amplitude H_d in Co_2MnSi 1) for ^{59}Co spin echo for symmetric (▲) and asymmetric (◻) influence at. $\tau_1 = \tau_2$ = 2 μs, $\Delta\tau$ = 10 μs, τ_d = 3 μs, f_{NMR} =145 MHz., and 2) for ^{55}Mn for symmetric (Δ) and asymmetric (■) influence at: τ_1 = 0.8 μs, τ_2 = 0.9 μs, $\Delta\tau$ = 8 μs, τ_d = 1.6 μs, f_{NMR} =353 MHz. I_o – echo mplitude at H_d=0,[1].

The dependence of its intensity on the time of application of a MVP for cobalt thin magnetic film (TMF), and ^{59}Co echo signal in Co_2MnSi (Figure 1*b* and 5*a*), correspondingly, differs considerably from closer-to-each-other dependences for lithium ferrite (Figure 1*a*) and ^{55}Mn echo signal dependence in Co_2MnSi (Figure 2*b*). The spectral diagrams of MVP influence of TMF under the influence of MVP are given in Figure 1*c*, which correspond to the symmetrical and asymmetrical actions of MVP. As can be seen from Figs. 1 and 2, the MVPs noticeably affect the echo spectra.

Note the difference in the TD of ^{59}Co and ^{55}Mn in Co_2MnSi. This indicates the difference in the anisotropy of HFF for these nuclei

So, the type of TDs is defined mainly by the hyperfine field anisotropy of corresponding nuclei which is small for ^{57}Fe and ^{55}Mn as compared for ^{59}Co positions.

The reason for this it could be understood from Figure 2c, where reduced echo intensity dependences on the MVP amplitude is shown for ^{55}Mn and ^{59}Co positions in Co_2MnSi. It is seen that for these positions amplitude dependences for asymmetric action of MVPs differ much stronger than ones for symmetric influence.

While a TD type is defined mainly by the anisotropic part of hyperfine interaction the degree of suppression of echo signals by MVP strongly depends on the DW mobility.

Note also that TDs in case of half metallic Co_2MnSi give visual pictures describing different hyperfine anisotropies at the ^{59}Co and ^{55}Mn sites.

In the cases we studied, the type of the TD was determined by the anisotropy of HFF and did not change for the given nucleus with increasing mobility of DW in the material (for instance, for polycrystalline cobalt and TMF of cobalt).

We can qualitatively understand the origin of two types of TD from the form of amplitude dependences of the MVP effect in Co_2MnSi, Figure 2c. The forms of dependences of echo signals on the symmetric and asymmetric effects for two different locations in one sample of Co_2MnSi show how the movement of the same DWs affects the echo signals from ^{59}Co and ^{55}Mn nuclei, which essentially differ in the anisotropy of HFF. It is clear that the diagrams of the MVP effect are close in the symmetric excitation case and differ essentially in the asymmetric effect case. These peculiarities can be understood if we assume that the diagrams of the symmetric effect are mainly determined by the mobility of DWs, i.e. by the coercive force of the sample, whereas the diagrams of the asymmetric MVP effect are mainly determined by the anisotropy of HFF. Hence we can now understand the dependence of the echo signal under the symmetric MVP effect with application of the external magnetic field H_0=1000 Oe in lithium ferrite, Figure 3. The obtained result testifies that the remained DWs are fixed in stronger pinning centers with application of the external field and hence that the echo signals are less suppressed under the symmetric MVP effect.

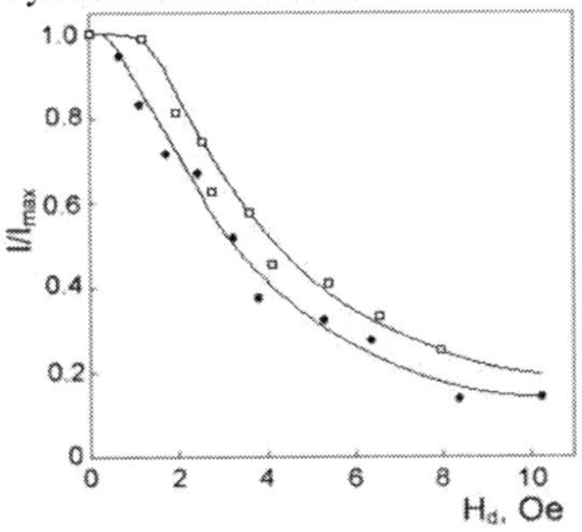

Figure 3. Amplitude dependence of the TPE intensity at MVP H_d, H_0=0 Oe (•) and H_0=1000 Oe (□)

Different behaviors of these dependences at different values of H_0 are clearly expressed in Figure 3. To achieve the same intensity of TPE at H_0=0 and H_0=1000 Oe, different values of H_d are necessary.

As was mentioned above, recording the dependence of echo signal intensity on the amplitude H_d ($I(H_d)$), it is possible to determine quantitatively the value of the MVP amplitude causing the shift in the DW equal to its effective thickness. This fact allows us to assess easily and quantitatively the DW pinning and the coercive force of the sample for the nuclei contributing the intensity of the NMR spin echo signal.

Figure 4 shows the experimental data on the dependences of the echo signal intensity on the MVP amplitude H_d under the symmetric MVP effect for cobalt nanopowder with mean grain diameter 100 nm, synthesized by electron beam technology [2], cobalt nanopowder with grain diameter 22 nm, produced by Sun Innovations, Inc., USA, and polycrystalline cobalt powder, grain size ~50 μm.

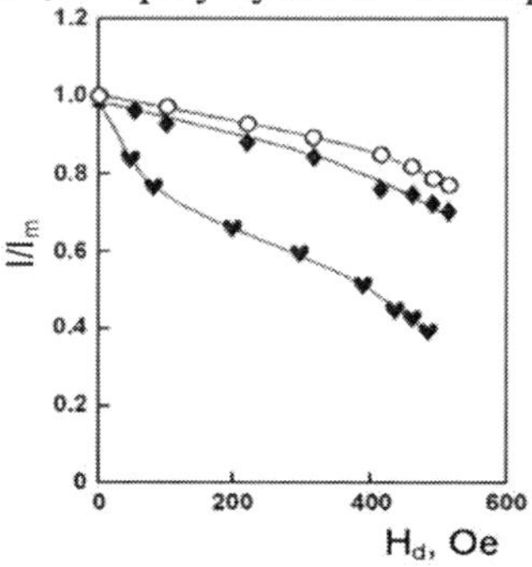

Figure. 4. Dependences of the echo intensity on the MVP amplitude H_d under the symmetric MVP effect for cobalt nanopowder with average grain diameter ~22 nm (○) and ~100 nm (♦) and coarse-grained powder with grain size ~50 μm (♥).

As is well known, a decrease in the grain size of cobalt powder from the order of 1μm to ~150 nm is accompanied by a significant increase in the coercive force of the sample associated with the decreasing size and the increasing part of surface effects. This fact also reflects on the dependence of the echo intensity on the value of the MVP. In all samples the echo signal intensity hardly changed up to the magnetic field values of the order of an average demagnetizing field because of the presence of DW. However, in the case of nanopowders, the mobility of DW decreases significantly, which should reflect on the character of the $I(H_0)$ dependence of the echo intensity on the value of the external magnetic field H_0. Besides, because of the increase in the coercive force, a much more powerful external field will be needed to suppress the echo signal in nanopowders. The measurement results on the dependence of the echo signal intensity on the value of the external magnetic field support the above reasoning, Figure 5.

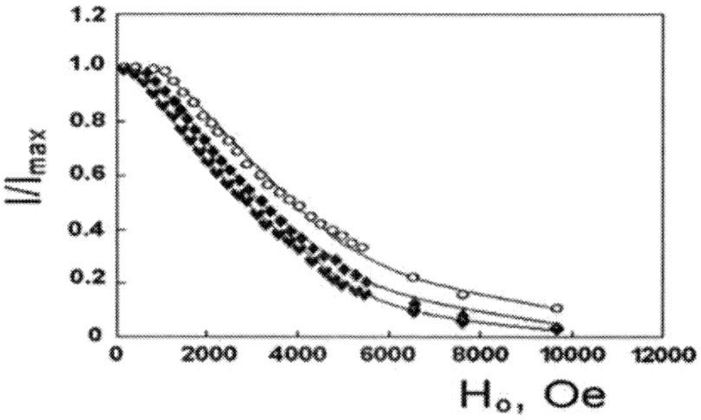

Fig. 5. Dependences of the NMR echo intensity on the value of the external magnetic field for Co powders: Co nanopowder, average grain diameter 100 nm, synthesized by electron beam technology (♦); Co nanopowder, average grain diameter 22 nm, produced by Sun Innovations, Inc., USA, (○); coarse-grained cobalt powder, grain size ~50 μm (♥)

The DW pinning can also increase when the grain sizes of cobalt nanopowders have decreased after their treatment with the help of a nanomill. In Fig. 6 the dependences of the echo signal intensities on the value of the external magnetic field are shown for polycrystalline cobalt powder with mean grain size ~50 μm and carbon nanopowder doped with cobalt nanoclusters (mean size - ~50 nm) [3] , and for the latter nanopowder after its grinding with a nanomill for half an hour, similar to the dependences shown in Figure 5.

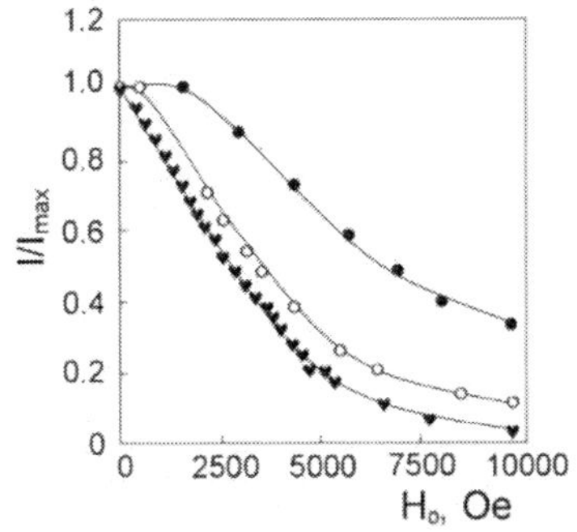

Figure 6. Dependences of the echo intensity on the value of the external magnetic field for Co powder with grains ~50 μm in size (♥) and carbon nanopowder doped with cobalt nanoclusters with grains ~50 nm in size (○), and for the latter nanopowder after grinding with a nanomill for half an hour (●)

A similar effect should be expected in the result of annealing of the sample.

For instance, Figure 7 shows the results of similar investigation for TMF cobalt samples fabricated in the form of films ~2 μm thick by the electroless deposition on quartz plates and annealed at 500°C for 3-4 hours [4]. They were also considerably magnetically softer than the cobalt powder with grains ~50 μm in size.

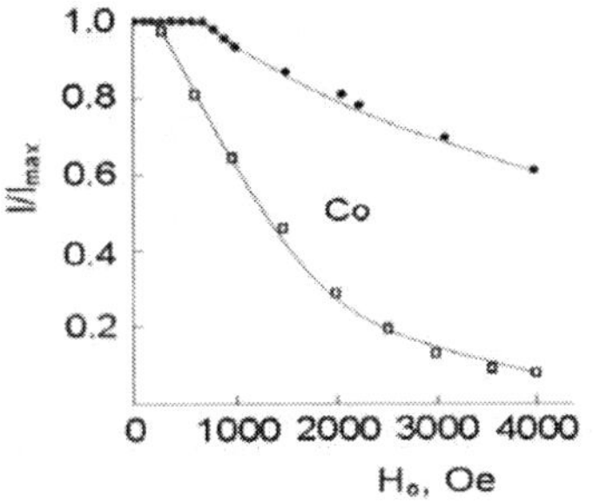

Figure 7. Dependence of the echo signal intensity for the cobalt film on the value of the external magnetic field before (●) and after (□) annealing.

Thus, based on the investigation of echo intensity dependence on the value of the external magnetic field, magnets can be classified by the degree of their magnetic rigidity or by the mobility of DW. For instance, in Fig.8 it is seen that, of half metal Co_2MnSi, cobalt and lithium ferrite samples, the magnetically softest sample is lithium ferrite, while half metal Co_2MnSi is the magnetically hardest one.

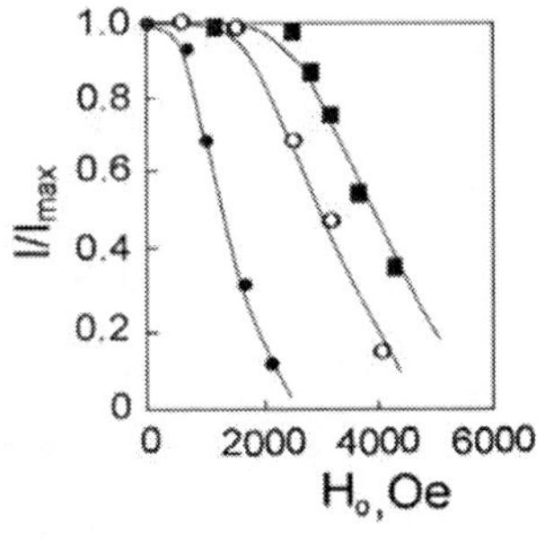

Figure 8. Dependence of the NMR echo intensity for 55Co nuclei in Co2MnSi (■), in cobalt (○), and for 57Fe in lithium ferrite (●) on the value of the external magnetic field.

It should be noted that, using the technique of chemical vapor deposition (CVD) [3], we produce carbon powders doped with magnetic nanoclusters with average diameter ~50 nm that have the coercive force of different degree depending on the temperature in the reactor chamber. The fabricated magnetic nanopowders could be both relatively magnetically soft and magnetically harder.

In Figure 9 it is shown the amplitude diagrams of the symmetric MVP effect for the nanopowders prepared at T=1200°C and 700°C.

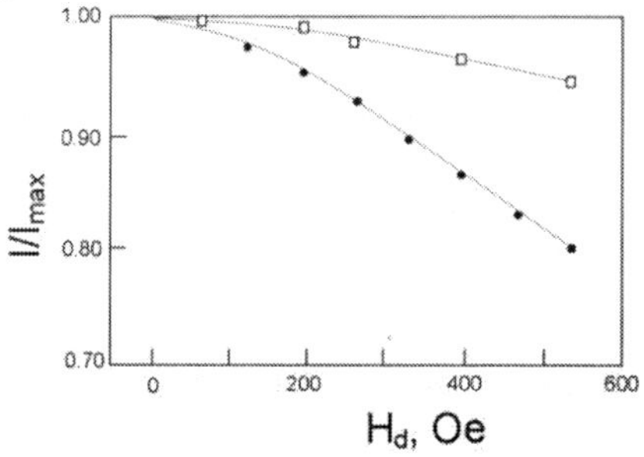

Fig. 9. Dependence of the nuclear spin echo intensity on the external magnetic field of carbon nanopowders doped with cobalt produced by the method of CVD at 1200°C (□) and 700°C (•)

At the same time, the variation in the degree of magnetic softness of the sample, i.e. the decrease in the DW pinning does not change the character of TDs of the MVP effect, because they are determined by the value of anisotropy of HFF of the nuclei.

Conclusion

In the result of experimental investigation of different types of TDs of the MVP effect, the existence of TDs of two types for the magnets under study was revealed. The type of TD is mainly determined by the anisotropy of HFF. We also studied the mechanism of the impact of the size of magnetic powder grains, and the pinning and mobility of domain walls on the revealed specific features of two-pulse echo signals observed under the MVP effect.

Acknowledgments

This work is supported by the Science and Technology Center in Ukraine (STCU) and the Shota Rustaveli National Science Foundation (SRNSF) Targeted Initiative Program 6081 grant.

References

[1] G.I. Mamniashvili, T.O. Gegechkori, A.M. Akhalkatsi, T.A. Gavasheli, E.R. Kutelia, L.G. Rukhadze, D.I. Gventsadze. Timing and spectral diagrams of magnetic video-pulse excitation influence on NMR spin-echo in magnets. 2013, *J. Supercond. Nov. Magn.*, **26**, 4, pp. 1401-1404.

[2] G.I. Mamniashvili, T.O. Gegechkori, M. Okrosashvili, E. Kutelia, A. Akhalkatsi, T. Gavasheli, D. Daraselia, D. Japaridze, A. Shengelaya, A. Peikrishvili, D. Lesuer. Production of cobalt nanopowders by electron-beam technology and their NMR and magnetometry study. 2015, *J. Magn. Magn. Mater*, **373**, pp. 177-182.

[3] T.O. Gegechkori, G.I. Mamniashvili, E. Kutelia, L. Rukhadze, N. Maisuradze, B. Eristavi, D. Gventsadze, A. Akhalkatsi, T. Gavasheli, D. Daraselia, D. Japaridze, A. Shengelaya. Technology for production of magnetic carbon nanopowders doped with iron and cobalt nanoclusters. 2015,*J. Magn. Magn. Mater.*, **373**, pp. 200-206.

[4] T.N. Khoperia. Electroless deposition in nanotechnology and ULSI. 2003, *Microelectronic Engineering*, **69**, 2-4, pp. 384-390.